IMPROVING ASSESSMENT AND EVALUATION STRATEGIES ON ONLINE LEARNING

ICLI is an annual International Conference on Learning Innovation (ICLI) hosted by Universitas Negeri Malang, Indonesia in collaboration with the Islamic Development Bank (IsDB) and Indonesian Consortium for Learning Innovation Research (ICLIR) as well as Univerisiti Teknologi MARA Cawangan Perlis, Malaysia serving as co-organizer this year. The conference aims to gather researchers, practitioners, students, experts, consultants, teachers and lecturers to share their insights and experiences on research not only in constructing innovations in learning but also the knowledge of learner's capability. The learners who are characterized as creative and competent by having the ability to understand what they have learned and capable of taking initiative and thinking critically. In addition, ICLI is organized on the basis of the trend in the 21st century, categorized by the increasing complexity of technology and the emergence of a corporate restructuring movement.

This book is the proceeding of ICLI 2021, containing a selection of articles presented at this conference as the output of the activity. Various topics around education are covered in this book and some literature studies around specific topics on learning and education are covered as well. This proceeding book will be beneficial to students, scholars, and practitioners who have a deep concern in education. It is also futuristic with a lot of practical insights for students, faculty, and practitioners, and also a description of the Indonesian educational system in today's era.

PROCEEDINGS OF THE 5TH INTERNATIONAL CONFERENCE ON
LEARNING INNOVATION (ICLI 2021), MALANG, INDONESIA, 29 JULY 2021

Improving Assessment and Evaluation Strategies on Online Learning

Edited by

Surjani Wonorahardjo
State University of Malang, Indonesia

Sari Karmina
Department of English, State University of Malang, Indonesia

Habiddin
Universitas Negeri Malang, Indonesia

Routledge
Taylor & Francis Group

Routledge is an imprint of the Taylor & Francis Group, an informa business

Typeset by MPS Limited, Chennai, India

Library of Congress Cataloging-in-Publication Data

A catalog record has been requested for this book

First published 2022
Published by: CRC Press/Balkema
Schipholweg 107C, 2316 XC Leiden, The Netherlands
e-mail: enquiries@taylorandfrancis.com
www.routledge.com – www.taylorandfrancis.com

ISBN: 978-1-032-19890-3 (Hbk)
ISBN: 978-1-032-19894-1 (Pbk)
ISBN: 978-1-003-26134-6 (eBook)
DOI: 10.1201/9781003261346

Improving Assessment and Evaluation Strategies on Online Learning – Wonorahardjo, Karmina & Habiddin (eds)
© 2022 copyright the Editor(s), ISBN: 9781032198903
Open Access: www.taylorfrancis.com, CC BY-NC-ND 4.0 license

Table of contents

Preface vii
Acknowledgement ix
Organizing Committee xi
Scientific Committee xiii

Improving expressive communication in children with autism using social story and picture exchange communication system 1
E. Kenila, E. Ediyanto, M. Efendi & I. Hitipeuw

Utilization of interactive multimedia to improve the reading ability of students with mild intellectual disabilities 8
N. Yuniarti, E. Ediyanto, M. Efendi, A. Sunandar, A.R. Junaidi & D. Hartanto

Innovation for early childhood based on augmented reality 16
D.R. Rini, A.M. Wisesa, R.T. Wulandari & T.F. Prihandini

Development of Android-based interactive multimedia on interaction of living things with the environment topic for seventh grade junior high school to improve student's learning motivation 23
Munzil, Y. Affriyenni & K. Kharismaliyansari

U-learning: Rethinking assessment for 21st-century learners 29
M.F. Yahaya

Development website of planning, writing, and publication of scientific articles based on Classroom Action Research (CAR) to increase teacher's pedagogical competence 37
H.R. Widarti, Habiddin, N.C.E. Setiawan, Parlan, A.B. Syafruddin, A. Ardyansyah & D.A. Rokhim

Implementation of the discovery learning model in regional economics 44
F. Rahmawati, I. Mukhlis & E. Yusida

Religiosity and nomophobia among undergraduate student: The moderating role of self-control 49
N.N. Afifah & H.E. Wijaya

Compromising peer assessment using Google Form in an online essay-writing course 56
N. Ariani & Y. Febrianti

Student self-regulated learning strategies during Covid-19 pandemic 63
N. Eva & S. Andayani

Experience using a five-component blended learning strategy during the Covid-19 pandemic 68
N.C.E. Setiawan & M.S. Rosli

Conceptual development of helper's assessment skills training design on inmate's resilience in L-SIMA 75
N. Hidayah, M. Ramli, L. Fauzan, M. Amin & H. Hanafi

"She is a support system": Peer assessment in cooperative learning asynchronous classroom 82
S. Karmina, A.H. Sujiatmoko, L. Setyowati, B.I. Lestari & W. Xiong

The effectiveness of contextual problem–solving-based integrative online learning on fundamental physics for higher education 89
Y. Affriyenni, S. Sutopo & C.I. Yogihati

Improving vocational education quality through life-based learning 96
Yoto, P.A.N. Mawangi & A.N. Pramudhita

Investigating the memory retention in extension concept mapping 101
D.D. Prasetya, A. Pinandito, Y. Hayashi & T. Hirashima

Development of virtual reality-based learning media on chemical bond materials and molecular shapes for grade 10th of senior high school students 108
A. Febriana, R. Joharmawan, R. Hakiki & M. Muchson

The Kirkpatrick model integration as an evaluation design for the implementation of vocational educator training in the 21st century 115
E. Sutadji, D.E. Rahayu, N. Fadilah, T. Sukmasari, A.B.N.R. Putra & A.M. Rini

Identification of intermolecular force misconceptions in students with different scientific thinking skills and the improvement efforts using conceptual change text 123
D.M. Ma'rufah, Effendy & S. Wonorahardjo

Improving the quality of online chemistry learning – A systematic literature review 129
I.W. Dasna & P.A.W. Putri

STEM-PBL and its effect on improving students' concept understanding in high school chemistry learning 135
P.A. Wulandari, I.W. Dasna & Nazriati

Author index 141

Improving Assessment and Evaluation Strategies on Online Learning – Wonorahardjo, Karmina & Habiddin (eds)
© 2022 copyright the Editor(s), ISBN: 9781032198903
Open Access: www.taylorfrancis.com, CC BY-NC-ND 4.0 license

Preface

The Industrial Revolution 4.0 and Society 5.0, the Covid-19 pandemic era and its disruptive technology require innovation in education and learning, where now evaluation and more are of critical importance. This is the era of a new type of challenges, also for higher education institutions likethe State University of Malang. This is also the time to prepare the young generation to shape the future of our society. Many approaches and innovative methods should be suitable for the generation, the educational resources should respond to the global issues. Universitas Negeri Malang (UM), through the Institute of Education and Learning Development (LP3), has been working to meet the requirements of this modern era. This will also bring the involvement of trans-discipline approaches where the disruptive technology issues will be addressed. UM has also initiated the centre of excellence (CoE) for Learning Innovation, The PUI-PT Disruptive Learning Innovation, with the support of the Ministry of Research, Technology, and Higher Education. This is an annual collaborative academic event organized by UMin collaboration with the Islamic Development Bank (IsDB), Center of Excellence Universitas Jember (UNEJ), Universitas Mulawarman (UNMUL), and Universitas Sultan Ageng Tirtayasa (UNTIRTA). This year we enhance our collaboration with an overseas university: Univerti Teknologi MARA (UiTM) Cawangan Perlis Kampus Arau.

This International Conference on Learning Innovation 2021 was designed for the experts of education and learning, to come together to make efforts to respond to the disruptive technology worldwide. With the spirit of cooperation and collaboration, it indeed could provide better comprehension of our situation, as well as preparing a guideline for modern education which covers all modern issues. The conference was held online at the State University of Malang, East Java, Indonesia on July 29, 2021. This Conference was organized by the Institute of Education and Learning Development, State University of Malang. There were five keynote speakers, six invited speakers, and one special speaker in this conference. They were part of the educational expertise around the globe. Prof. Gavin Brown (University of Auckland, New Zealand), Prof. Stuart Kime (Evidence Based Education, United Kingdom), Prof. Weishen Wu, PhD (Dayeh University, Taiwan), Dr. Eng. Muhammad Ashar, S.T., M.T. (State University of Malang, Indonesia), and Dr. Mohammad Fadhili Yahaya (University Teknologi Mara Perlis Branch, Malaysia) were the keynote speakers. In addition, we have one special speaker Dr. Rer. Nat. Suseno Amien (Learning Innovation, PMU IsDB, Indonesia) and four invited speakers from the Center of Excellence 4 in 1 IsDB and two invited speakers from UiTM, Malaysia. They were Dr. Eng. Didik Dwi Prasetya, S.T., M.T. (UM), Dr. Rida Oktorida Khastini (UNTIRTA), Dr. Latisha Asmaak Shafie (UiTM), Anton Rahmadi, S.TP., M.Sc., Ph.D (UNMUL), Erlia Narulita, S.Pd., M.Si., Ph.D. (UNEJ), dan Dr. Razlina Razali and Dr. Farah Lina Azizan (UiTM):.

The organizers wish to acknowledge the keynote speakers for their presentation on ICLI 2021, also the services provided by the reviewers of the papers for their time, hard work, and dedication to this Conference. In addition, many thanks to all persons who helped and supported this conference. Furthermore, we would also like to invite presenters around the world to participate in the next ICLI that will be held at the State University of Malang. Finally, we hope that the future event will be as successful as ICLI 2021, as indicated in this proceeding.

Editor,

Surjani Wonorahardjo, PhD

Acknowledgement

The editor and organizer of the International Conference on Learning Innovation, ICLI 2021 wish to acknowledge the keynote speakers for their presentation on ICLI 2021. They are Prof. Gavin Brown (New Zealand), Prof. Stuart Kime (United Kingdom), Prof. Weishen Wu, PhD (Taiwan), Dr. Eng. Muhammad Ashar, S.T., M.T. (Indonesia), Dr. Mohammad Fadhili Yahaya (Malaysia), Dr. Rer. Nat. Suseno Amien (Islamic Development Bank, IsDB, Indonesia), Center of Excellence Universitas Jember (UNEJ), Univesrsitas Mulawarman (UNMUL), and Universitas Sultan Ageng Tirtayasa (UNTIRTA) and Universitas Negeri Malang (UM), our co-organizer Universiti Teknologi MARA (UiTM) Cawangan Perlis Kampus Arau, all sponsors, both organizing and scientific committee members, and all participants for making this conference possible and successful.

The organizers also wish to acknowledge publicly the valuable services provided by the reviewers. On behalf of the editors, organizers, authors, and participants of this Conference, we wish to thank the keynote speakers and the reviewers for their time, hard work, and dedication to this Conference. Without their services, the editors could not maintain the standards of Learning Innovation in this digital age.

The organizers wish to apologize to the speakers who could not publish his/her paper in this Conference Proceeding. Our apology also given to all participants for all shortcomings in this conference. Hopefully, we will see you in the next ICLI at the Universitas Negeri Malang

Malang, January 15, 2022
Eli Hendrik Sanjaya, S.Si., M.Si.,Ph.D.
Chairperson ICLI 2021

Improving Assessment and Evaluation Strategies on Online Learning – Wonorahardjo, Karmina & Habiddin (eds)

Organizing Committee

Advisory Board
Prof. Dr. Ah. Rofi'uddin
Universitas Negeri Malang, Indonesia

Prof. Ts. Dr. Mohd Azlan Mohd Ishak
Universiti Teknologi MARA Cawangan Perlis, Malaysia

Chairman:
Assoc. Prof. Eli Hendrik Sanjaya, S.Si., M.Si., PhD.
Universitas Negeri Malang, Indonesia

Secretary
Assist. Prof. Sari Karmina, S.Pd., M.Pd., PhD.
Universitas Negeri Malang, Indonesia

Treasurer
Assoc. Prof. Ella Faridati Zen, M.Pd
Universitas Negeri Malang, Indonesia

Technical Chair
Assist. Prof. Andika Bagus Nur Rahma Putra, S.Pd, M.Pd.
Universitas Negeri Malang, Indonesia

Assoc. Prof. Dr. Razlina Razali
Universiti Teknologi MARA Cawangan Perlis, Malaysia

Publication
Assoc. Prof. Surjani Wonorahardjo, PhD
Universitas Negeri Malang, Indonesia

Prof. Dr. Latisha Asmaak Shafie
Universiti Teknologi MARA Cawangan Perlis, Malaysia

Improving Assessment and Evaluation Strategies on Online Learning – Wonorahardjo, Karmina & Habiddin (eds)
© 2022 copyright the Editor(s), ISBN: 9781032198903
Open Access: www.taylorfrancis.com, CC BY-NC-ND 4.0 license

Scientific Committee

Improving Assessment and Evaluation Strategies on Online Learning – Wonorahardjo, Karmina & Habiddin (eds)

Improving expressive communication in children with autism using social story and picture exchange communication system

E. Kenila, E. Ediyanto*, M. Efendi & I. Hitipeuw
Universitas Negeri Malang, Malang, Indonesia

ABSTRACT: In everyday human activities, both social and personal interrelations require communication. Cognitive abilities influence the communication process. Cognitive disorders in children with autism lead to communication problems. Communication disorders in children with autism lead to emotional changes. In the learning process, children with autism can overcome communication disorders by maximizing their visual potential. In this article, we study that a review of the visual media literature can improve visual abilities of children with autism. The method used is a narrative review with steps to 1) determine the scope, 2) search literature, and 3) analyze the findings. The results found that to improve communication skills and basic behavior in children with autism, the Picture Exchange Communication System (PECS) can be used. PECS can get used to communicating with children with autism by using symbols that are appropriate to their daily activities. To increase the effectiveness, the implementation of PECS can be combined with the Social Story Model. Social stories can help students with autism interpret and understand social situations with therapy using short stories. By using PECS and SS, student with autism can increase their positive social behavior and the development of receptive and expressive communication. It also can improve structured speech skills and stimulate language.

Keywords: Student with Autism, Picture Exchange Communication System, Social Story, Expressive Communication

1 INTRODUCTION

Communication has a vital role in human life, both socially and in personal relationships. Because it is so important for humans, scientists try to answer every difficulty so that effective communication can occur by creating various communication media as it is today. Communication impairment slows development in people by many ways. Communication itself is an activity that receives and transfers information in the form of ideas, messages through speech, images, symbols, writing, or behavior (Troshanska & Trajkovski 2014). As social beings, humans need other people to complement each other and learn many aspects. To complement each other, effective and intense communication and relationships are needed (Rueda 2020). Good communication occurs strongly influenced by cognitive abilities and experience where a person can formulate a message so that it can be understood by the communicant so that he or she can do the right things as conveyed by the message and actualize it according to the concept and context (Hallahan, Kauffman & Pullen 2020).

Social communication impairment is the main disorder in children with autism, repetitive behavior, and limited interests (Hyman, Levy, & Myers 2020). This communication disorder causes autistic children to face several difficulties, such as difficulty in understanding simple instructions and difficulty expressing what they want or feel. Not only autistic children, teachers, parents, or

*Corresponding Author

DOI 10.1201/9781003261346-1

caregivers also have difficulty understanding what autistic children want. As a result, they are often easily sad or throw tantrums because what they want or need cannot be conveyed. On the other hand, other people are also unable to understand the message delivered clearly. Thus, there is a gap between children with autism and their environment (Cagliani et al. 2017; Hyman, Levy, & Myers 2020). If not treated early, this problem will worsen both children with autism and the people around them.

Although having a disorder in social communication, Koshino et al. (2008) states that a group of children with autism can process letters visually, in a non-verbal way, and not verbally. So the children with autism are also referred to as visual learners (Tissot & Evans 2003). Children with autism can compensate by utilizing visual learning to help children with autism in building and developing communication, incredibly expressive communication, such as social stories and Picture Exchange Communication System (PECS) (;ondy & Frost 2011; Riga, Ioannidi, & Papayiannis 2020).

Based on the explanation above, the current study wants to analyze whether social stories and PECS can improve expressive communication in children with autism based on literature studies. Currently, many methods have been developed to help improve communication in children with autism.

2 METHODS

The method used in writing this article is a narrative review, which summarizes several research results and compares them to produce a natural or holistic interpretation of children with autism found in academic articles (academic-oriented literature). The stages carried out by the researchers in this narrative review consist of three stages (Gasparyan et al. 2011), including 1) determining the scope of focus and the results of the research. The primary approach studied in this scientific research is to find the focus of research concerning social stories media or the PECS method. The step of the selection of the article can be seen in Figure 1. The results produced in this scientific search are in terms of finding the effect of social stories, media, and PECS on developing expressive communication in children with autism. 2) Conducting a literature search through keywords related to the title of this research. The search for published articles was conducted on Google Scholar and Research Gate with the keywords "PECS and/or social stories to improve expressive communication in children with autism." The articles determined are in 2011–2021, while the criteria for the journals reviewed are 16 articles.

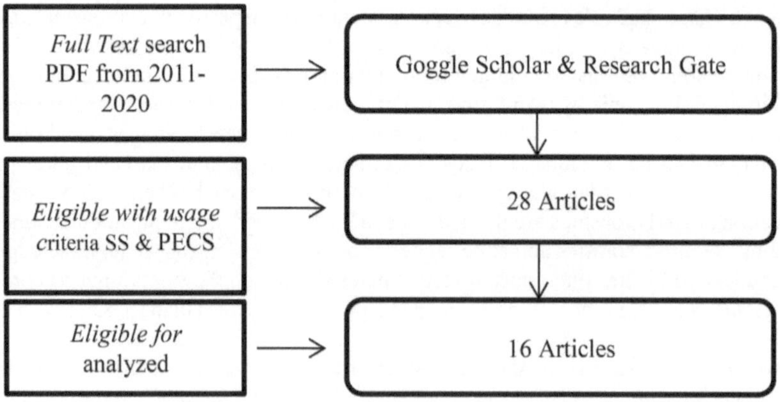

Figure 1. Article selection scheme.

The third stage is to write the findings from the article. The collected article is analyzed and summarized in Table 1. The summaries of these articles are then discussed against the other relevant studies.

3 RESULTS

Bondy and Frost use applied behavior analysis to understand and improve verbal behavior. They used communication function analysis according to antecedents or triggers and their consequences. The analysis shows that communication is a function of the consequence of his behavior. Evidence presented throughout his book show that the use of communication systems other than language enhances speech. Communication systems development provides other significant benefits, such as reducing frustration due to the current unavailability of effective alternative communication strategies (Bondy & Frost 2011). So the use of the PECS method has been proven to be effective in improving expressive communication in children with autism (Asep et al. 2019; Goa & Teresia 2018; Putri, Hastuti, & Adi 2018). PECS can also improve structured speech skills (Vistasari & Patria 2019) and stimulate spoken language or verbally among students his age (Jusoh & Majid 2017).

Furthermore, Paul (2012) underlined the value that a PECS image and visual communication strategy in children with autism could improve receptive and expressive communication as a parallel process with the same ability. Some children have good receptive abilities but lack expressive skills or vice versa. Therefore, parents and caregivers during the lockdown are advised to use PECS as a medium of daily communication. The house is an important part of lives. There is no need to wait for a particular time for communication exercises (Bondy, Horton, & Frost 2020). However, Purnama, Safitri, and Zwagery (2019), on the contrary, found that the PECS method could not significantly improve communication skills in children with autism.

Table 1. Summarized research that focuses on PECS and SS in students with ASD.

Author(s)	Research Focus	Findings
Sunandar et al. (2019).	The test is the effectiveness of visual aid methods (PECS and Visual Schedule) on children's communication skills and basic behavior	The use of visual support (PECS & Visual schedules) can improve the communication skills and behavior of children with autism
Putri, Hastuti, & Adi (2018).	The effect of the PECS method toward the communication ability of autistic children	The conclusion showed that the PECS method affects the communication ability of autistic children
Vistasari, & Patria (2019).	To increase structure speaking in children with autism using PECS	The results showed an increase in structured speaking skills in both participants through the PECS program that had been carried out
Purnama, Safitri, & Zwagery (2019).	Improvements in communication skills in children with autism by PECS method	The PECS method cannot significantly improve communication skills in autistic children
Goa, & Derung (2017).	To determine the expressive communication skills of children with autism using the PECS	The results obtained showed an increase in children's expressive communication with autism after using the PECS method
Bondy, Horton, & Frost (2020).	Promoting functional communication skills within homes	Describes nine essential communication skills and provides examples of how families can improve the use of these important skills
Paul, Kalyanpur, & Harry (2012).	The value of a PECS image and visual communication strategy in autistic children	Communication has additional subcomponents, namely expressive and receptive. The evidence presented throughout the book shows that communication systems other than language improve speech development and provide other significant benefits, such as reducing frustration over the current availability of effective alternative communication strategies

(continued)

Table 1. Continued.

Author(s)	Research Focus	Findings
Jusoh, & Majid (2017).	This study aimed to see the improvement in pronunciation of words by autistic students by using PECS	PECS is successful in stimulating speech among students
Ling (2017)	Social stories concepts, goals, functions, and guidelines for creating social stories guide for kids	Social stories are ready to be used according to the situation and problems of each child
Da Silva, Arantes, & Elias (2020).	Social stories in class affect presenting social stories in appropriate learning and reducing inappropriate behavior of children with ASD in the hall	Change is reducing challenging behavior and increasing positive behavior Social stories represent strategies designed in reducing inappropriate behavior and improving adequate in the classroom, as a short implementation time and in supervising the construction of the two materials used
Meister (2020).	Evaluating the effectiveness of social stories for autistic children	The social story method is more effective in reducing challenging behaviors, and the results are more variable when used to improve the behavior to be developed or social skills
Aldabas (2019).	Examine the effectiveness of technology integration when using social stories	Social stories are effective in reducing inappropriate social behavior among these children. Several studies have been about applied social stories with the use of technology
Garzotto, et al (2018).	Education for subjects with neurodevelopmental disorders (NDD) to help them gain an accurate understanding of social situation interventions, develop autonomy, and learn appropriate behavior	Exploiting Wearable Immersive Virtual Reality (WIVR) technology is called creating a new form of the social story called Wearable Immersive Social Story (WISS) This study was designed to control for the learning benefits of wearable immersive social stories and traditional social stories.
Daneshvar, Charlop, & Berry Malmberg, (2019).	To compare the effectiveness of the two procedures, the Photo Activity Schedule Intervention, and the Social Stories Intervention, in teaching social skills to four children with Autism Spectrum Disorders (ASD).	All of these support the use of interventional photography activities to teach social skills to children with ASD. We discuss the inconsistent findings of the effectiveness of social stories
Balakrishnan, & Alias (2017).	Use of social stories to encourage social interaction in children with autism spectrum disorders	The findings of this study indicate that the use of social stories to increase the social interaction of children with ASD has a positive impact
Riga, Ioannidi, & Papayiannis (2020).	A brief overview of the use of Social Stories (SS) as a strategy to support students in achieving communicative skills	This paper argues that using SS as an invaluable teaching method that can be integrated into the teaching and learning process, especially for children with ASD, special educators use SS and thus can contribute to children acquiring and developing social and communicative skills

The social story method also showed effectiveness in change, reducing challenging behavior, and increasing positive behavior (Da Silva, Arantes, & Elias 2020). Social skills also positively impact (Aldabas 2019; Balakrishnan & Alias 2017; Meister 2020) social interaction and communication (Riga et al. 2020). Therefore, Ling (2017) pays attention to social stories by formulating concepts, functions, and guidelines for making social stories. Several topics or themes can be used according to each child's situation, problems, and needs. Furthermore, Garzotto, Gelsomini, Matarazzo, Messina, & Occhiuto (2018) utilize Wearable Immersive Virtual Reality (WIVR) technology to create a new form of the social story called Wearable Immersive Social Story (WISS). This media is used for educational interventions for children with autism to help them gain an accurate understanding of social situations, develop independence, and learn to adapt to new behaviors. This study was conducted to determine the benefits of learning WISS compared to more traditional social stories tailored to each person's specific needs with NDD. However, the findings of Daneshvar, Charlop, & Berry Malmberg (2019) are highly controversial, which compared the effectiveness of two procedures, the Activity Schedule intervention in the form of Photographs and the Social Stories Intervention, in teaching social skills to children with disabilities. "Autism Spectrum Disorders (ASD)" stated that support for its findings is inconsistent concerning social stories. More details about the articles are summarized in Table 1.

4 DISCUSSION

One of the main symptoms of children with autism is difficulty in social communication. The kind of social communication problems include difficulty understanding what is happening in their environment, difficulty understanding instructions, difficulty expressing wants and needs, difficulty using and understanding language so that they often use behavior as a form of communication to express basic needs. The same thing was also conveyed by Skinner (1957). Paul (2012) said that all behaviors are crying, hurting yourself, snatching, eye contact, smiling, and using pictures or symbols, and others, which are as a result of communication. The judgment is needed to find the trigger or antecedent "what caused the behavior to occur?", "why did this behavior occur?". and "how to solve it?".

PECS is one of the media and a method for analyzing behavior and understanding and developing language or verbal behavior more positively. PECS was first initiated to change complex behavior and form positive behavior to develop expressive communication using symbols or verbal in children with autism, sentence structure, and language stimulation. PECS consists of six phases that show the process of practicing the communication of these behaviors to form and develop positive behavior, which is communication, both verbally and or through images with a better structure if able to reach that stage. Each phase in PECS is adjusted to the progress of each child and the stimulus is given. Moreover, then children understand what to do if they want to get what they like. This exercise is based on Skinner's theory of operant conditioning (Bondy & Frost, 2010 in Thiemann 2020).

A social story is a teaching media that can be integrated into the learning process at school and home, especially for children with autism. So special education schools can use social media stories to develop social and communication skills (Riga, Ioannidi, & Papayiannis 2020). It can only be done if children with autism have successfully passed the PECS protocol. Meanwhile, listening to a good understanding requires a good one. At the same time, receptive communication is the basis for development of expressive communication (Breitenbach, Armstrong, & Bryson 2013). The social stories using technology can be a choice for behavior and communication development in educational interventions so that children can gain an accurate understanding of social situations, development of independence, and adjustment of new behaviors.

The Aided Language Stimulation (ALS) is also used to develop vocabulary and sentences in children with autism. This method also provides benefits for developing receptive communication in understanding words through pictures, objects, and actions by imitating communicators according to relevant situations simultaneously. So, the facilitator and communicant combine Augmentative and Alternative Communication in realistic and concrete situations. The teacher or communication partner points to the picture on the communication sheet and stimulates with words. At the same time, movement or modeling is relevant to the activity (Acheson 2016). In addition, the strategies to teach communication is using visual support based on activities in a contextual environment (Fitzgerald 1994 in Dada & Alant 2009).

5 CONCLUSION

Both the PECS method and the social story media have the same properties, namely the formation and development of positive social behavior and the development of receptive and expressive communication in parallel, improving structured speech skills, stimulating language or oral students in parallel their age class. Good social stories are used before the child passes the PECS protocol. Social stories refer to the effectiveness of development, both behavior and social communication. Augmentative and alternative communication (AAC) in ALS is also a communication medium used to develop receptive and expressive communication. At the same time, technology-based social stories can be used as an alternative to shape and develop social behavior and communication in children with autism. Children with autism required routine and intense personal interrelation so that social behavior and communication are formed and patterned correctly through PECS as a holistic and natural strategy and approach.

The PECS approach is recommended to be used to develop pre-cursor skills on a regular and consistent basis. Before passing the PECS protocol, use social media stories to develop social behavior and communication to develop on an ongoing basis according to the problems and needs. To develop more effective expressive communication, people can use ALS and technology-based social stories as alternative media.

REFERENCES

Acheson, M. 2006. The Effect of Natural Aided Language Stimulation on Requesting Desired Objects or Actions in Children with Autism Spectrum Disorder. A Dissertation Submitted to the Division of Graduate Education and Research of the University of Cincinna.

Aldabas, R. (2019). Effectiveness of social stories for children with autism: A comprehensive review. Technology and Disability, 31(1–2), 1–13. https://doi.org/10.3233/TAD-180218

Balakrishnan, S., & Alias, A. (2017). Usage of Social Stories in Encouraging Social Interaction of Children with Autism Spectrum Disorder. Journal of ICSAR, 1(2), 91–97.

Breitenbach, M. M., Armstrong, V. L., & Bryson, S. E. (2013). The implementation of best education practices for a student severely affected by autism. International Journal of Inclusive Education, 17(3), 277-294.

Bondy, A., & Frost, L. (2011). A picture's worth: PECS and other visual communication strategies in autism. Woodbine House.

Bondy, A., Horton, C., & Frost, L. (2020). Promoting Functional Communication Within the Home. Behavior Analysis in Practice, 13(2), 321–328.

Cagliani, R. R., Ayres, K. M., Whiteside, E., & Ringdahl, J. E. (2017). Picture Exchange Communication System and Delay to Reinforcement. Journal of Developmental and Physical Disabilities, 29(6), 925–939. https://doi.org/10.1007/s10882-017-9564-y

Da Silva, M. C., Arantes, A., & Elias, N. C. (2020). Use of social stories in classroom to children with autism. Psicologia Em Estudo, 25, 1–15.

Dada, S., & Alant, E. (2009). The effect of aided language stimulation on vocabulary acquisition in children with little or no functional speech, 18(1), 50–64.

Daneshvar, S. D., Charlop, M. H., & Berry Malmberg, D. (2019). A treatment comparison study of a photo activity schedule and Social Stories for teaching social skills to children with Autism Spectrum Disorder: brief report. Developmental Neurorehabilitation, 22(3), 209–214. https://doi.org/10.1080/17518423.2018.1461947

Garzotto, F., Gelsomini, M., Matarazzo, V., Messina, N., & Occhiuto, D. (2018). Designing wearable immersive "Social stories" for persons with neurodevelopmental disorder. In Lecture Notes in Computer Science (including subseries Lecture Notes in Artificial Intelligence and Lecture Notes in Bioinformatics): Vol. 10907 LNCS. Springer International Publishing. https://doi.org/10.1007/978-3-319-92049-8_37

Gasparyan, A. Y., Ayvazyan, L., Blackmore, H., & Kitas, G. D. (2011). Writing a narrative biomedical review: considerations for authors, peer reviewers, and editors. Rheumatology international, 31(11), 1409.

Goa, L., & Derung, T. N. (2017). Komunikasi Ekspresif Dengan Metode Pecs Bagi Anak Dengan Autis. Jurnal Nomosleca, 3(2). https://doi.org/10.26905/nomosleca.v3i2.2037

Hallahan, D. P., Pullen, P. C., Kauffman, J. M., & Badar, J. (2020). Exceptional learners. In Oxford Research Encyclopedia of Education.

Hyman, S. L., Levy, S. E., & Myers, S. M. (2020). Identification, Evaluation, and Management of Children With Autism Spectrum Disorder. Pediatrics, 145(1), 1–71.

Jusoh, W., & Majid, R. A. (2017). Using Picture Exchange Communication System To Improve Speech Utterance Among Children With Autism. Journal of ICSAR, 1(1), 46–49.

Koshino, H., Kana, R. K., Keller, T. A., Cherkassky, V. L., Minshew, N. J., & Just, M. A. (2008). fMRI investigation of working memory for faces in autism: visual coding and underconnectivity with frontal areas. Cerebral cortex, 18(2), 289–300.

Ling, J. (2017). Social Stories for Kids in Conflict. In Social Stories for Kids in Conflict.

Meister, K. (2020). An Evaluation of the Effectiveness of Social Stories for Children with Autism.http://search.proquest.com/openview/3a02417f9d5d2a8d23aa92d190b457a7/1?pqorigsite=gscholar & cbl = 18750 & diss=y

Paul, H. A. (2012). Kalyanpur, M., & Harry, B.(2012). Cultural Reciprocity in Special Education: Building Family–Professional Relationships. Baltimore, MD: Paul H. Brooks.

Purnama, M. A. A., Safitri, J., & Zwagery, R. V. (2020). Penerapan Metode Picture Exchange Communication System (PECS) Dalam Meningkatkan Keterampilan Berkomunikasi Pada Anak Dengan Autisme. Jurnal Kognisia: Jurnal Mahasiswa Psikologi Online, 2(2), 83–86.

Putri, C. R. R., Hastuti, W. D., & Adi, E. P. (2018). The Influence the Picture Exchange Communication System Method toward The Communication Ability of Autistic Child. Journal of ICSAR, 2(2), 180–185.

Riga, A., Ioannidi, V., & Papayiannis, N. (2020). Autism Spectrum Disorder and Social Stories Purpose and Utility. Asian Journal of Education and Social Studies, 12(4), 19–26.

Rueda, C. (2020). Neuroeducation: Teaching with the brain. Journal of Neuroeducation, 1(1), 108–113.

Sunandar, A., Indreswari, H., Efendi, M., Dian Firdiana, A., & Lailiyah, N. (2019). The Use of Visual Support of PECS and Schedule Based on the Characteristics of Student with Autism. 382(Icet), 559–564. https://doi.org/10.2991/icet-19.2019.140

Thiemann, J. (2020). Didactic concept of teaching material development. In the development of a haptic simulation game for the area of procurement and the investigation of the learning effect (pp. 59–112). Springer Gabler, Wiesbaden.

Tissot, C., & Evans, R. (2003). Visual teaching strategies for children with autism. Early Child Development and Care, 173(4), 425–433.

Troshanska, J., & Trajkovski, V. (2014). Stages of communication in children with autism. 3(1), 71–80.

Vistasari, R., & Patria, B. (2019). Program PECS (Picture Exchange Communication System) untuk Meningkatkan Kemampuan Berbicara Terstruktur pada Anak Autis. Gadjah Mada Journal of Professional Psychology (GamaJPP), 5(1), 94.

Improving Assessment and Evaluation Strategies on Online Learning – Wonorahardjo, Karmina & Habiddin (eds)

Utilization of interactive multimedia to improve the reading ability of students with mild intellectual disabilities

N. Yuniarti, E. Ediyanto*, M. Efendi, A. Sunandar & A.R. Junaidi
Universitas Negeri Malang, Indonesia

D. Hartanto
Universitas Ahmad Dahlan, Indonesia

ABSTRACT: The current study is a literature study that describes interactive multimedia in learning to read for students with mild intellectual disabilities. The research used written sources from various peer-reviewed journal articles of good quality. Articles are selected according to the research topic collected from the Google Scholar database. Researchers reviewed 15 journal articles published in the 2015–2020 period with the topic of the role of interactive multimedia in learning to read for students with mild intellectual disabilities. The study results show that interactive multimedia, such as Kinect, Computer Assisted Instruction, Marbel mobile application, and pop-up books can increase the effectiveness of learning and reading skills for students with mild intellectual disabilities. In addition, interactive multimedia can support interesting interactions and a fun learning experience between teachers, students, and learning.

Keywords: interactive multimedia, reading ability, student with mild intellectual disabilities.

1 INTRODUCTION

Reading is one of the language skills used to communicate by getting information from something written, including letter recognition activities, translating symbols into meaning, word recognition, and understanding the meaning of reading (Aulina 2012; Bayu 2018; Munthe & Sitinjak 2019). Therefore, reading is an essential step in a child's development. According to Fauziah (2016), by mastering reading skills, children have the basis for mastering other fields of study. If they cannot read at an early age, the child will have difficulty learning the field of study in the next class. From this statement, it can be concluded that reading skills are very important for all children, including children with intellectual disabilities. According to Purnamasari & Soendari (2018), learning to read in children with intellectual disabilities becomes difficult because of the impact of their intelligence barriers, namely being unable to think abstractly, lacking concentration, experiencing memory, and perception disorders.

Santrock (2017) argues that intellectual disabilities is a condition with symptoms before the age of 18 that involves low intelligence and difficulties adapting to everyday life. They have disorder experienced during the developmental period that includes deficits in intellectual and adaptive functions in the conceptual, social, and practical domains (Vicario & Hernández 2013). Therefore, experts can conclude that children with intellectual disabilities are children who have a level of intelligence below normal, so that to carry out the development tasks they need special assistance or services, including in the field of education or academics. According to Atmaja (2018), in terms of education or academics, children with intellectual disabilities experience obstacles in several areas,

*Corresponding Author

 DOI 10.1201/9781003261346-2

namely attention, memory, language development, low motivation, and their academic abilities, which are below average for their age. One of the impacts of these obstacles is having difficulty in learning to read.

To overcome the problems above, we need learning media (assistive devices) to help mild children with intellectual disabilities improve their reading skills. Developments in the world of education today using learning media that utilize interactive multimedia are not foreign to support conventional learning methods. Interactive multimedia combines various media in text, images, graphics, sound, animation, video, and interaction packaged into digital files (computerized) to convey messages or subject matter to students (Atmawarni 2011). The use of interactive multimedia in the teaching and learning process is very effective and efficient in delivering reading learning.

This study analyzes interactive multimedia learning to read for children with intellectual disabilities based on literature studies in line with the various descriptions above. This study is important considering the development of technology that leads to interactive multimedia as an effective and efficient learning medium in improving the reading ability of mild children with intellectual disabilities.

2 METHODS

The current study provided a clear overview of the available research. A literature review was conducted by selecting articles from journals in the article selection process carried out in a peer-review stage. The selected journal articles are publications from 2015 to 2020. The steps taken in this research are field selection, analysis, and categorization. Relevant articles have been selected in this study that focus on the theme of Using Interactive Multimedia in Learning to Read for Mild Intellectual Disabilities Students.

The selection of articles starts from research journals that apply peer-review on Google Scholar. Before searching for keywords, filter only articles with peer review, and the full text is available. The keyword used in the search was "Interactive Multimedia To Improve Learning, Mentally Impaired." From this search, 360 articles were found. After screening the first article, namely articles published from 2015 to 2020, 50 articles were found. In the second screening, articles containing research questions were selected, "Utilization of Interactive Multimedia in Learning to Read for Mild Intellectual Disabilities Students." We also used "Mental Retardation" term for "Intellectual Disabilities." Fifteen articles were identified for thorough analysis according to research questions.

Each article is further reviewed, keeping in mind some of the research questions that have been set. Finally, the researcher ensures that each article used as a reference is in accordance with the research question and provides answers to the research questions.

The literature review was carried out by utilizing the best evidence approach from the articles found. Based on 15 articles consisting of five international articles and 10 national articles, it was in accordance with the research question.

3 RESULTS

The advancement and use of technology in education is a fundamental need to support the effectiveness and quality of learning. The use of technology to develop learning media is very much needed for children with intellectual disabilities in understanding the material presented by the teacher, especially in reading. Learning media developed by utilizing ICT can be audio, visual, audiovisual, 3D images, animated videos, games, and multimedia (Nurcholis & Nur 2018; Sagirani 2015). They must pay attention to the characteristics and needs of students in developing ICT-based learning media for children with intellectual disabilities. The media must be accessible to operation and availability of applications (easy to obtain), take advantage of an attractive appearance, not easily damaged, not abstract, and use sensory modes as much as possible to strengthen student learning

(Ambarwati & Darmawel 2020; Fauziah 2016; Fauzia & Kustiawan 2017; Nurcholis & Nur 2018; Ratna 2019; Sagirani 2015). When planning learning using computer assistance to provide instructional reading for students with mild intellectual disabilities, teachers need to consider cognitive accessibility and cognitive demands (Burt et al. 2020).

Cognitive barriers in children with intellectual disabilities cause delays in capturing knowledge, storing, and processing analytical knowledge. So we need learning media that can provide direct experience in learning to read. For this reason, learning media for reading in interactive CDs is developed, which is equipped with exciting pictures, audio, video, and animation. It turns out that interactive CD media can encourage children with intellectual disabilities to study independently, which can increase students' motivation and concentration to positively impact the results of early reading learning abilities (Fauzia & Kustiawan 2017; Ratna 2019).

Other interactive multimedia used in learning to read are those that can be installed on a computer or laptop and can display animated videos in early reading learning for children with intellectual disabilities or kindergarten level. Research conducted by Diwangkara et al. (2016) and Lalu Satya and Edy (2019) proves that interactive multimedia can awaken children's enthusiasm and involvement in learning to read, a very positive and significant influence on early reading skills. Another study conducted by Burt et al. (2020) using computer-assisted instruction (CAI) has proven to be a viable way to teach vocabulary to students with intellectual disabilities.

Table 1. The interactive multimedia in learning for student with intellectual disabilities.

Author(s)	Title	Finding(s)
Hall et al. (2015)	Addressing learning disabilities with UDL and technology: Strategic reader	• CAST created a technology-based system that combines Universal Design for Learning (UDL) and Curriculum-Based Measurement (CBM) in a digital learning environment to improve reading comprehension instruction for students with learning needs The research also shows that technology can help teachers do their jobs more effectively
Sagirani (2015)	Utilization of **Kinect** in Learning Media Application Prototypes for Children with Special Needs	• Media development with Kinect can support the formation of exciting interactions and fun learning experiences for children with intellectual disabilities in learning to read Kinect is a device that can capture and decipher body movement and sound. Existing audio and video information serves as commands for interacting with digital content presented in software programs. Kinect has several essential characteristics, among others. First, Kinect is a flexible tool in its use, meaning that a teacher (teacher) can interact with students through body movements and voice without using input devices such as keyboards and mice. Second, Kinect can accommodate some of the needs and interests of the interaction of more than one user. The three Kinect are tools that can collect 3D (three-dimensional) information so that they can support teaching activities that have many variations
Diwangkara et al. (2016)	Development of 3 Dimensional Interactive Learning Media Illustrated Reading for Children with Intellectual Requirements	• The advantage of Computer Assisted Instruction (CAI) for children with intellectual disabilities is that the computer can accommodate students who are slow to accept learning because it can provide a more effective climate more individually, never forgets, never gets bored, is very patient in carrying out instructions as desired by the program used 3 Dimensional Interactive Learning Media Illustrated Reading for Mentally Impaired Children in Triamerta Singaraja Kindergarten Based on Computer-Assisted Instruction designed based on a predetermined Thematic Syllabus, and using Use Case Diagrams and Activity Diagrams as functionality and flow of activities in the system

(continued)

Table 1. Continued.

Author(s)	Title	Finding(s)
Fauziah (2016)	Use of Intelligent Interactive Multimedia Learning to Read Beginning in Mild Mentally Impaired Children	• Interactive multimedia for intelligent learning is a learning media published by Gramedia which contains materials for beginning reading. Through this interactive multimedia, children can learn to read the beginning with the word symbol method, where children will learn to read words accompanied by pictures of the word The use of intelligent interactive multimedia for learning to read affects improving the early reading ability of mild children with intellectual disabilities
Fauzia & Kustiawan (2017)	Interactive Multimedia to Improve Beginning Reading Skills of Mentally Impaired Students	• The interactive multimedia product I love to read has been adapted to the characteristics of children with intellectual disabilities and is made attractively and to make it easier for students to use. This media is equipped with a manual. The teacher or student assistant will understand the function and how to use this media. This media is equipped with attractive images, audio, video, and animation. This media uses the SAS method in delivering reading material that is equipped with audio. This media is in the form of a CD (compact disk), making it easier to use and store. This media can encourage students to learn independently.
Nurcholis & Nur (2018)	Application of Marbel Letters on the Ability to Recognize Letters for Class II Mild Mentally Impaired Children at SLB Wonogiri	• Marbel Letters mobile application learning media is effective for the ability to recognize letters, among others, because, in Marbel Letters, there are games/games and exciting materials for children Marbel Letter media can be used for letter recognition and introduce colors to children. This way makes Marbel an efficient media because it contains more than one material and saves costs. In one technology, it is multifunctional and can be used for various benefits.
Komalasari et al. (2019)	Interactive Multimedia Based on Multisensory as a Model of Inclusive Education for Student with Learning Difficulties	• Multisensor-based interactive multimedia is a learning media that can accommodate various types of student learning by optimizing the different senses possessed by students, namely auditory, visual, kinesthetic, and tactile By using multisensor-based interactive multimedia, students are more enthusiastic and motivated to listen to learning materials.
Rini Djuwita Ratna (2019)	Improving Reading and Writing Skills for Beginning Mentally Impaired Students in Class D3-C Using ABACADA Interactive CD Media at SLB-C Negeri Pembina South Kalimantan	• The use of ABACADA Interactive CD electronic media can motivate children with intellectual disabilities to learn, read, and write. ABACADA Interactive CD electronic media is a multifunctional teaching aid. In addition to learning to read and write, it also functions as a playing medium that can improve concentration and fine motor (fine motor). This media is liked by students because it uses interesting pictures and can be moved with a mouse. This media provides learning centered on student activities based on individuality, where each individual uses their media that can increase children's motivation and concentration in learning
Afriyanti & Ardisal (2019)	Improving Word Reading Skills Through Educational Game Media for Mentally Impaired Children at SLB Perwari Padang	• The ability to read words in mild children with intellectual disabilities can be improved through educational game media. It is proven that the provision of intervention using educational game media Educational game media are games accompanied by learning and are the latest learning media that can improve understanding quickly because they are supported by interesting games and make students active.

(*continued*)

Table 1. Continued.

Author(s)	Title	Finding(s)
Ahmad & Yin (2019)	Using interactive media to support reading skills among underachieving children	• Children with special needs will benefit from interventions in word recognition, phonological awareness, and reading comprehension performance using interactive multimedia tools to help develop their future reading skills. Educators need to move towards the use of customized and modified interactive multimedia as an instructional tool in teaching and learning, thereby improving the reading skills performance of students with special needs from the presence of multimedia in education
Lalu Satya & Edy (2019)	Application of IT-Based Video Media on the Ability to Read Simple Sentences for Class III Mild Mentally Impaired Children	• The use of IT-based media for children with intellectual disabilities can improve their ability to read simple sentences. In addition, IT-based media can also awaken children's enthusiasm for reading, as well as full involvement of students in learning
Gökbulut & Güneyli (2019)	Printed Versus Electronic Texts in Inclusive Environments: Comparison Research on the Reading Comprehension Skills and Vocabulary Acquisition of Special Needs Students	• Using assistive technology such as computers, tablets, and overhead projectors in the classroom will help increase the effectiveness and permanence of learning activities, especially for students with special needs Electronic texts are more effective in improving vocabulary skills and reading comprehension for students with special needs than printed texts
Ambarwati & Darmawel (2020)	Implementation of Multimedia Development Life Cycle in Learning Media Applications for Children with Intellectual Requirements	• Interactive learning media can help teachers increase the activity & understanding of children with intellectual disabilities Learning with adobe flash game applications on computers using adobe flash software is well received by children with intellectual disabilities and provides great potential for them in learning reading and mathematics
Idhartono (2020)	The Effectiveness of Pop Up Book Media on the Reading and Writing Ability of Mild Mentally Impaired Students in SLB	• Media pop-up book is a book that has pictures as if it can move and is in three-dimensional form and can create good communication to readers through visuals, textual, images, and writing. In addition, the pop-up book involves its readers to be more active and interactive in the discussions contained in the book so that it is easy to understand something being discussed. With pop-up book media, children with intellectual disabilities can imagine and interact with what they read by touching the pictures that appear in the book. Parents and teachers will also find it easier to teach children to read because the media that children will read attracts their hearts
Burt et al. (2020)	Computer-assisted instruction (CAI) has proven to be a viable way to teach vocabulary to students with intellectual disabilities	• Computer-assisted instruction (CAI) has proven to be a viable way to teach vocabulary to students with intellectual disabilities. CAI is a broad term for the use of structured instructional strategies through electronic or multimedia interfaces that partially or wholly replace teachers

Another form of interactive multimedia used to read for children with intellectual disabilities is Pop-Up Book media. The physical composition and content of the book in the form of three dimensions on the pop-up book media creates a very high interest and interest for children with intellectual disabilities to learn the material. From research conducted by Idhartono (2020), students can easily understand the content of the material being studied because the three-dimensional images contained in the book seem to move, causing interest in learning and increasing students' ability to learn reading comprehension material. In addition, the pop-up book media also involves students to be more active and interactive through visuals, textual, images, writings, and discussions contained in the book.

The results of other studies that have been carried out show that the use of learning media in educational games can improve the ability to recognize letters and read words in mild children with intellectual disabilities. The intervention was carried out by providing teaching using educational game media and evaluation using an action test. Educational game learning media can increase understanding quickly because it is supported by interesting games and makes students active (Afriyanti & Ardisal 2019; Nurcholis & Nur 2018).

In developing a technology-based system that combines Universal Design for Learning (UDL) and Curriculum-Based Measurement (CBM) in a digital learning environment to improve reading comprehension instructions for students with learning disabilities, it is necessary to pay attention to the following principles: (1) provide several ways of representation to anticipate and overcome first the physical, perceptual, and cognitive barriers that may interfere with student learning; (2) provide multiple ways of action and expression by recognizing the variability that students use to plan, strategize, perform learning tasks; and (3) provide multiple ways of engagement, namely emphasizing that students must be involved or motivated to learn (Hall et al. 2015).

4 DISCUSSION

The use of digital media in learning for early childhood (3–6 years) has overgrown at this time. Learning with digital media is delivered using computers and online video games, mobile devices, electronic toys, dolls, or robots that move and talk. The use of digital media has been researched and shows effectiveness in children's learning. Many digital media products and interactive toys are designed to learn to read, such as the alphabet, phonics, and word recognition. Research has shown that well-designed digital learning shows effective results. Digital media can also improve vocabulary, spelling, and reading skills better than traditional teaching methods. In addition, online learning has helped children learn foreign languages, and internet-based digital stories will also increase understanding of foreign languages. Learning with digital media also provides opportunities for expression and imagination in domains such as music, drawing, painting, story-telling, and poetry. Computer learning activities can generate a high level of interest and focus on learning tasks that do not diminish over time. Attention to computer-based learning can be exceptionally high among preschoolers, including in some cases, those with attention deficit hyperactivity disorder (ADHD). It can happen because the computer can provide immediate feedback on the child's performance and progress (Lieberman et al. 2009).

Children with intellectual disabilities need IT-based learning media to improve their cognitive abilities. Improving cognitive abilities in children with intellectual disabilities can use various methods and media that attract children's interest to learn, one of which is interactive multimedia. Learning that uses interactive multimedia in which audio and visual elements can train children's thinking and memory, especially for children with mild intellectual disabilities, is a suitable choice because the material presented is concrete unconventional. Using interactive multimedia can improve the quality of learning because the material and presentation of learning are very interesting. Multimedia is also very relevant to shorten teaching because the material presented will be delivered quickly compared to conventional learning. With multimedia, students also more easily understand the subject matter, especially interactive multimedia. One of the media that can be used in learning to read for children with intellectual disabilities is a video tutorial learning media with Adobe Flash software/application. It can be concluded that interactive multimedia, which includes audio and visuals, is very influential and helps deliver material by educators in the classroom because it creates fun learning, motivates, increases children's enthusiasm in learning. Such a situation can improve children's memory because learning materials are presented in a concrete form and packaged in a form that increasingly attracts students' attention. Thus, it can also be concluded that interactive multimedia has a significant effect on increasing the learning achievement of mild children with intellectual disabilities (Maulidiyah 2020).

Another form of interactive multimedia used in learning to read for early childhood students, including children with intellectual disabilities, is the educational game media. Educate games can provide insight or teaching that can make players or students practice their ability to run a game and facilitate students to reach their stage of development, one of which can facilitate children in reading skills. The use of educative games in learning is motivated by the development of technology and applications on smartphones. Most children, including children with intellectual disabilities, spend their time watching videos or playing games with smartphones and are less interested in reading books or stories. So the development of educative games to attract students' attention to use this media has a learning function to improve reading skills. Reading hobbies and skills will provide various perspectives and broader insights to children in various ways. It will develop children's creative thinking patterns, so they must be improved from an early age because children who like to read have more capable language skills. Parental participation is needed to assist students when using educative games and ensure that children will not access other applications that are not relevant to reading skills (Febriani et al. 2018).

The use of technology is one of the efforts to meet the future needs of students, so that collaboration between students and teachers is needed in creating learning that develops digital skills in achieving a predetermined learning goal. In addition, the teachers must pay attention to the accessibility of the technology to be used, principles of learning for students, such as when choosing to make video calls. Furthermore, they must pay attention to student attitudes and readiness and reinforce students' reinforcement and social manifestations in the community. In addition, before learning is carried out, plan to learn well and clearly according to the learning objectives to the evaluation tools. It will be selected and what will be achieved, pay attention to the characteristics of different students, and form cooperation with students' parents in helping access the website during learning (Ersanty 2020).

5 CONCLUSION

The application of interactive multimedia in learning in schools is one solution to create interesting and innovative learning. An interactive multimedia is a tool in conveying materials and learning objectives for mild children with intellectual disabilities. For this reason, interactive multimedia must deliver an attractive appearance in terms of images, letters, commands, navigation, and audio. Furthermore, the interactive multimedia used should be easy to understand for children with intellectual disabilities. There are many interactive multimedia options for learning. The teacher must pay attention to the characteristics of students with mild intellectual disabilities and needs that may differ from one another. In addition, collaborating with parents to assist students when using interactive multimedia in learning is an important aspect that must be considered.

This research has provided a reasonably clear picture of interactive multimedia learning to read for students with mild intellectual disabilities that provides effective results. It is hoped that teachers will continue to strive to develop learning media following information needs and technological developments so that learning activities are more active and more effective according to shared expectations by considering the characteristics and needs of students with special needs. For this reason, it is necessary to support school policies so that they continue to develop teacher human resources in the field of information and communication technology and improve facilities and infrastructure so that learning effectiveness can be achieved.

REFERENCES

Afriyanti, D., & Ardisal. (2019). Meningkatkan Kemampuuan Membaca Kata Melalui Media Game Edukasi Untuk Anak Tunagrahita di SLB Perwari Padang. *Journal of Multidicsiplinary Research and Development*, 2(1), 154–161.

Ahmad, N. A., & YIN, K. Y. (2019). Using interactive media to support reading skills among underachieving children. *International Journal of Innovation, Creativity, and Change*, 8(7), 81–88.

Ambarwati, P., & Darmawel, P. S. (2020). Implementasi Multimedia Development Life Cycle Pada Aplikasi Media Pembelajaran Untuk Anak Tunagrahita. *Majalah Ilmiah UNIKOM*, 18(2), 51–58.

Atmaja, R, J. (2018). *Pendidikan Dan Bimbingan Anak Berkebutuhan Khusus*. Bandung: PT Remaja Rosdakarya.

Atmawarni. (2011). Penggunaan Multimedia Interaktif Guna Menciptakan Pembelajaran Yang Inovatif Di Sekolah. *Perspektif*, 1(1), 20–27.

Aulina, C. N.(2012). Pengaruh Permainan Dan Penguasaan Kosakata Terhadap Kemampuan Membaca Permulaan Anak Usia 5-6 Tahun.*PEDAGOGIA:Jurnal Pendidikan*, 1(2),131–143.

Bayu, P. (2018). *Asesmen Membaca Permulaan Sebagai Upaya Deteksi Dini Anak Berkesulitan Belajar Membaca (Dyslexia)*. 1–15.

Burt, C., Graham, L., & Hoang, T. (2020). Effectiveness of Computer-assisted Vocabulary Instruction for Secondary Students with Mild Intellectual Disability. *International Journal of Disability, Development, and Education*, 1–22.

Diwangkara, I. K. S., Putrama, I. M., & Sunarya, I. M. G. (2016). Pengembangan Media Pembelajaran Interaktif 3 Dimensi Baca Tulis Bergambar Untuk Anak Tunagrahita. *Kumpulan Artikel Mahasiswa Pendidikan Teknik Informatika*, 5(1).

Ersanty, D. (2020). Pembelajaran Berbasis Web Untuk Siswa Tunagrahita Ringan. *Pendidikan Khusus*, 1–14.

Fauzia, A., & Kustiawan, U. (2017). Multimedia Interaktif untuk Meningkatkan Kemampuan Membaca Permulaan Siswa Tunagrahita. *Jurnal ORTOPEDAGOGIA*, 3, 6–12.

Fauziah, P. (2016). Penggunaan Multimedia Interkatif Cerdas Belajar Baca dalam Meningkatkan Kemampuan Membaca Permulaan pada Anak Tunagrahita Ringan (Studi Eksperimen Dengan Desain Single Subject Research Terhadap Siswa Tunagrahita Ringan Kelas III SDLB C Cinta Asih). *Jurnal Unik*, 1.

Febriani, A. S., Mulyana, E. H., & Rahman, T. (2018). Pengembangan Educative Game Berbasis Aplikasi Android Untuk Memfasilitasi Keterampilan Membaca Anak Usia 5-6 Tahun. *Jurnal Paud Agapedia*, 2(2), 187–196.

Gökbulut, Ö. D., & Güneyli, A. (2019). Printed versus electronic texts in inclusive environments: Comparison research on the reading comprehension skills and vocabulary acquisition of special needs students. *Education Sciences*, 9(3).

Hall, T. E., Cohen, N., Vue, G., & Ganley, P. (2015). Addressing learning disabilities with UDL and technology: Strategic reader. *Learning Disability Quarterly*, 38(2), 1–12.

Hidalgo Vicario, M. H., & Rodríguez Hernández, P. R. (2013). DSM-5. manual diagnóstico y estadístico de los trastornos mentales. últimas novedades. In *Pediatria Integral* (Vol. 17, Issue 7).

Idhartono, A. R. (2020). *Keefektifan Media Pop Up Book Terhadap Kemampuan Membaca dan Menulis Siswa Tunagrahita Ringan di SLB*. 2(1), 8–13.

Komalasari, M. D., Pamungkas, B., Wihaskoro, A. M., Jana, P., Bahrum, A., & Khairunnisa, N. Z. (2019). Interactive Multimedia Based on Multisensory as a Model of Inclusive Education for Student with Learning Difficulties. *Journal of Physics: Conference Series*, 1254(1).

Lalu Satya, P., & Edy, R. (2019). Penerapan Media Video Berbasis IT Terhadap Kemampuan Membaca Kalimat Sederhana Anak Tunagrahita Ringan Kelas III. *Pendidikan Khusus*,1–10.

Lieberman, D. A., Bates, C. H., & So, J. (2009). Young children's learning with digital media. *Computers in the Schools*, 26(4), 271–283. https://doi.org/10.1080/07380560903360194

Maulidiyah, F. N. (2020). Media Pembelajaran Multimedia Interaktif Untuk Anak Tunagrahita Ringan. *Jurnal Pendidikan*, 29(2), 93–100.

Munthe, A. P., & Sitinjak, J. V. (2019). Manfaat Serta Kendala Menerapkan Flashcard Pada Pelajaran Membaca Permulaan. *Jurnal Dinamika Pendidikan*, 11(3), 210–228.

Nurcholis, A. F., & Nur, A. (2018). Pengaruh Mobile Application Marbel Huruf Terhadap Kemampuan Mengenal Huruf Anak Tunagrahita Ringan Kelas II Di Slb Negeri Wonogiri. *Jpk: Jurnal Pendidikan Khusus*, 13(2), 41–56.

Purnamasari, P,.& Soendari, T. (2018). Volume 19 Nomor 1, Juni 2018. *Metode VAKT Untuk Pembelajaran Membaca Permulaan Anak Tunagrahita Ringan*, 19(1), 25–31.

Rini Djuwita Ratna. (2019). Meningkatkan Kemampuan Membaca Dan Menulis Permulaan Siswa Tunagrahita kelas D3-C Dengan Menggunakan Media CD Interaktif ABACADA Di SLB-C Negeri Pembina Provinsi Kalimantan Selatan. *JurnalIlmiah ProGuru*, 5(2),225–237.

Sagirani, T. (2015). Pemanfaatan Kinect dalam Prototype Aplikasi Media Pembelajaran bagi Anak Berkebutuhan Khusus. *Jatisi*, 2(1), 1–13.

Santrock, W, J. (2017). *Psikologi Pendidikan* (5th ed.). Salemba Humanika.

Improving Assessment and Evaluation Strategies on Online Learning – Wonorahardjo, Karmina & Habiddin (eds)

Innovation for early childhood based on augmented reality

D.R. Rini*, A.M. Wisesa & R.T. Wulandari
Universitas Negeri Malang, Indonesia

T.F. Prihandini
Universitas Negeri Yogyakarta, Indonesia

ABSTRACT: The current learning process requires teachers to be more innovative and creative in delivering learning materials. One of the innovations provided by the teacher in delivering learning materials is innovation in developing learning media used for teaching. Along with the online learning process to inhibit the spread of Covid-19, teachers are required to develop technology-based learning media. Nowadays technology is getting more advanced. Many technologies, such as smartphones, can be used to deliver learning materials. This study develops Augmented Reality-based learning media for early childhood that can be operated via an Android smartphone. In the learning media developed, students will learn to recognize the names of animals and explain the characteristics of each of these animals. The method used is the ADDIE model. Before being used, the developed learning media was validated and tested. Validation is done by media expert validation and material validation. The trials were conducted with limited trials and open trials. From the results of the trial, an analysis was carried out to see how far the effectiveness of the developed learning media was. The development product produces Augmented Reality-based Flash Cards to recognize Elephants, Lions, Monkeys, Zebras, and Giraffes. The flash card is then scanned on the Virtual Zoo application, which was developed using Vuforia and Unity software. From the results of the study, it was observed that 100% of students could recognize the name of the animal and 91% of students could explain the characteristics of the animal.

Keywords: Innovation, learning media, early childhood, Augmented Reality

1 INTRODUCTION

We are currently entering the digital era (Reflianto & Syamsuar 2019). All activities that were previously carried out by humans are now assisted by digital technology that can make it easier for humans to carry out their duties. The digital era is the terminology for a time where everything starts with technology (Syamsuar & Reflianto 2018). This also applies in the field of education, the learning process that was previously carried out directly by teachers and students, with the advancement of technology can now be accessed from anywhere (Senduk et al. 2016). Especially the Covid-19 pandemic condition has made the learning process online-based.

While facing the Covid-19 pandemic the learning process is shifted from conventional learning to online learning, raising several new problems. Some of these problems include 1) Teachers must possess knowledge and capabilities in using the latest technology to deliver learning materials 2) Teachers must be able to create innovative digital-based learning media so that students can understand learning materials well, and 3) Teachers must master what learning platforms they can use for the learning process. Learning media are essential in early education, especially in this

*Corresponding Author

 DOI 10.1201/9781003261346-3

digital age (Giganti 2007). Digital media plays a significant part in early childhood engagement although has its own controversy (Lydia Plowman 2012).

Based on the Early Education Curriculum (Yuliantina 2021), one of the methods to stimulate learning in children is to make them use their sensory reflexes. A zoo is usually a very good place to give them a sensory experience. However, during this pandemic, a visit to the zoo is shortly on hold to reduce the spread of Covid-19 among children as well as the animals. Considering this, it is necessary to develop learning media that can provide real learning experiences for students. For example, in early childhood, in the absence of a pandemic, the teacher could invite children to learn to recognize animals directly at the zoo. However, due to the pandemic, direct learning was hindered. To provide a solution to this problem, through this research, an Augmented Reality-based Virtual Zoo Learning Media was developed, as a means to learn to recognize animals and their characteristics through 3-dimensional animal shapes observed through flash cards. The development of learning media based on visual observation is expected to attract students' interest in learning optimally. Augmented Reality has been widely used in various levels of education such as higher-level education, secondary education (lower/upper secondary level), primary education, and in informal learning (Masmuzidin 2018). In Abenyega (2008), visualization also plays a huge role in building understanding for students about their core, such as identity, belonging as well as their learning experience. Thus, learning media that involves more observation or visual-based learning media are expected to help students who can understand the material well. Research by (Rosmiati 2020) also states that audio-visual-based learning media is one of the alternative learning media in conducting technology-based learning processes.

2 METHOD

Facing the Covid-19 pandemic, which caused the learning process to shift from conventional learning to online learning, this raises several new problems. Some of these problems include 1) Teachers must be able to have knowledge and capabilities in using the latest technology to deliver learning materials 2) Teachers must be able to create innovative digital-based learning media so that students can understand learning materials well, and 3) Teachers must master what learning platforms they can use for the learning process. Learning media are essential in early education, especially in this digital age (Giganti 2007). Digital media plays a significant part in early childhood engagement although has its own controversy (Lydia Plowman 2012).

Based on the Early Education Curriculum (Yuliantina 2021), one of the methods to stimulate learning in children is to make them use all of their sensory. So, zoo is usually a very good place to give them sensory experience. But during this pandemic a visitation to zoo is momentarily on hold to stop the COVID-19 from spreading to children as well as to the animals. Considering this, it is necessary to develop learning media that can provide real learning experiences for students. For example, in early childhood, in the absence of a pandemic, the teacher can invite children to learn to recognize animals directly at the zoo. However, due to the pandemic, direct learning cannot be done. To provide a solution to this problem, through this research, an Augmented Reality-based Virtual Zoo Learning Media was developed, as a means to learn to recognize animals and their characteristics through 3-dimensional animal shapes observed through flash cards. The development of learning media based on visual observation is expected to attract students' interest in learning optimally. Augmented Reality has been widely used in various level of education such as higher-level education, secondary education (lower/upper secondary level), primary education, and in informal learning (Masmuzidin 2018). In Abenyega (2008), visualization also plays huge role in building understanding for students about their core, such as identity, belonging as well as their learning experience. So that learning media that involve more observation or visual-based learning media are expected to be able to contain students who can understand the material well. Research by (Rosmiati 2020) also states that audio-visual-based learning media is one of the alternative learning media in conducting technology-based learning processes.

3 RESULTS

3.1 *Product development*

Virtual zoo is one of the learning media in the form of flash cards. This learning media is intended for early childhood. The developed Flash Card contains animal recognition material, which refers to one of the basic competencies of the PAUD curriculum, namely, theme 3.8. Get to know the natural environment (animals, plants, weather, soil, water, rocks, etc.). From this theme, the research team took the sub-material, namely introducing animals (Farida Yusuf 2018). The designed Virtual Zoo is a learning media developed so that students can recognize the animals in the flashcard images in three dimensions. Flash card learning media is designed with Augmented Reality technology so that the images can be studied in a three-dimensional way. Augmented Reality (AR) is a form of interactive technology that combines virtual technology with the real world (Risdianto et al. 2021). AR technology can convert two-dimensional into three-dimensional images, this concept was implemented by the research team in developing learning media in the form of images. Virtual Zoo learning media presents three-dimensional images of animals intended for children who study the learning media to observe animals in three dimensions. This is to cope with learning amid the Covid-19 pandemic which does not allow children to learn directly about animals in their environment, for example in the park or in the nursery. Thus, by using learning media that presents animal shapes in three dimensions, it is hoped to provide a good experience for children to learn the characteristics of animals, such as animal's, characteristics, foo, and environment.

The display of illustrations of various kinds of animals on flash cards is designed using adobe illustration software. The illustration design is used as a marker, and then entered into the vuforia software to create a development key. The development key that has been created later on in the vuforia software will see the unique value of the marker created. The marker value is indicated by the number of stars displayed. The unique value on the marker must be above two, for visualization of Augmented Reality products (Ismayani 2020). The following are the results of a marker scan trial from animal illustrations drawn by the research team:

Figure 1. Animal marker scan results.

From the results of the marker scanning trial above, it was found that the markers on zebras, lions, monkeys, and giraffes have a five-star rating and elephants have a three-star rating. This shows that these markers can be further developed as Augmented Reality. The display of markers on the developed flash card can be seen in the following figure:

Fig. 2
Elephant Marker

Fig. 3
Giraffe Marker

Fig. 4
Monkeys Marker

Fig. 5
Lion Marker

Fig. 6
Zebras Marker

After the marker development on the flash card is complete and the marker scan results are valid, the next process is making the application. The application design was created using Adobe Illustrator and Affinity Design. Applications were made using Unity Software. In the process of making the marker application that has been entered in the vuforia database, it is then exported to unity and will be developed into a three-dimensional form. The developed application can only be operated on android devices. The front page of the application contains illustrations of animal images and the name of the application "Virtual Zoo". On that page, there are two arrows to choose from, the first is choosing to start scanning animals on an Augmented Reality-based flash card and the second is an information page about using the application. The application display is as shown in the following figure:

Fig. 7
Front page of application

Fig. 8
Information page of application

3.2 *Product trial*

The trial of using the application was carried out at the UM Laboratory PAUD KB School. The Product trial was conducted on 10 PAUD students with an age range of 3–4 years. The use of Flash Cards and the Virtual Zoo application is accompanied by their respective parents. To facilitate the understanding of parents and teachers in assisting students using the application, the research team also created a manual. Before the trial process was carried out, the research team explained the use of application features, ranging from application installation, flash card scans, to how to explain content to children. The flash card is made in 6.35 × 8.89 cm size, using 150-g art paper. In its development, we apply the Elam (2017) ergonomic principle, wherein the product proportion must comply with the user's body. The Virtual Zoo flash card that we have developed will be operated by parents with their children.

From the test results we obtained the following data:

No.	Indicators	Number of children	
1	The number of children who can recognize the name of the animal they are observing		
	Animal Name	Can Mention	Can't Say
	Elephant	10	0
	Zebra	10	0
	Monkey	10	0
	Giraffe	10	0
	Lion	10	0
2	The number of children who can mention the characteristics of the animals they observed		
	Animal Name	Can Mention	Can't Say
	Gajah	10	0
	Zebra	7	2
	Monyet	10	0
	Jerapah	10	0
	Lion	8	2

From the results of the experiments above, it was obtained that 100% of the total samples tested could name the animals they were observing. This shows that the visualization of each type of animal on the Flash Card Learning Media is easy for children to understand and recognize. The second indicator, after observing the Augmented Reality-based Virtual Zoo learning media, children can learn to recognize animals by looking at their shapes in three dimensions. From the test results, 91.83% of children can recognize the characteristics of animals. For example, they can mention the characteristics of a giraffe as an animal with a long neck. But they are still confused in recognizing the character of a lion, this is because they are confused with other animals such as cats or tigers.

In addition to getting data on student responses in recognizing the names and characters of animals in the trial activities, the research team also received positive feedback from parents and teachers. Parents and teachers are very enthusiastic about accompanying children to learn about animals using the Virtual Zoo and Augmented Reality-based Flashcard applications. According to them, the innovation of Virtual Zoo learning media and Augmented Reality-based Flash Cards is very good for sharpening children's imagination in recognizing animals. According to teachers, the use of digital application-based learning media is very necessary at this time, so they strongly agree with the development of Augmented Reality-based applications. This is in line with research (Risdianto et al. 2021) which states that the existence of Augmented Reality-based learning media strongly supports the learning process of blended learning.

Through this product, children can recognize animals in 3 dimensions by observing how many legs they have, what their anatomy is like, and they can also observe the animals they are studying from various angles. This provides an experience similar to when they visit a zoo and observe animals directly. This is in line with the research that has been done by (Ailsa Salsabila 2020), which states that by using Augmented Reality-based learning media for early childhood students can be more motivated and can make it easier to understand lessons. In addition, students get direct learning experiences through Augmented Reality-based learning media.

The trial application display can be seen in the following figure:

4 DISCUSSION

From the results of the development of Augmented Reality-based Flash Card learning media and Virtual Zoo applications, it is considered as an important contribution to KB Laboratory UM. During the Covid-19 pandemic, the learning process only takes place online through platforms Zoom, Whatsapp, Google Meet, etc. In introducing various types of animals, teachers and parents only display two-dimensional learning media. The development of Augmented Reality-based learning media can provide an experience for students to learn types of animals and can observe them in three dimensions.

With the development of digital-based learning media, it is also very good to train students and parents to use their smartphones for the learning process. The Covid-19 pandemic has caused the learning process to be carried out online (Sadikin & Hamidah, 2020). Therefore, the proper use of smartphones will be beneficial in the child's learning process (Handarini et al. 2020).

5 CONCLUSION

From this research activity, it can be said that the development of Flash Card learning media and Augmented reality-based Virtual Zoo applications is very good to help the learning process during the pandemic. The digital-based learning media can provide a real experience for students to learn various types of animals. The learning media that was developed was responded well and enthusiastically by the task people, teachers, and students. AR has passed the hype stage and can now be implemented for educational purposes (Sinduningrum et al. 2019). Augmented reality technology is able to improve student learning experiences and help identify animal species.

To improve the quality of the digital-based learning process, this research produces several recommendations for further research. For further research, the same Augmented Reality-based learning media can be developed with different content, for example with the theme of learning body parts, learning transportation tools, or learning various fruits and vegetables.

REFERENCES

Agbenyega, J. S. (2008). Developing The Understanding Of The Influence Of School Place On Students' Identity, Pedagogy And Learning, Visually . *International Journal Of Whole Schooling*, 52–66.

Ailsa Salsabila, C. (2020). Pembelajaran Menggunakan Augment Reality Untuk Anak Usia Dini Di Indonesia. *Jurnal Teknologi Pendidikan*, 5(1), 20–37. Retrieved from

Elam, L. (2017). *Ergonomic guidelines for designing handheld products a case study of hand-held vacuum cleaners*. 16–21.

Farida Yusuf. (2018). *Kerangka Dasar dan Struktur Kurikulum 2013 PAUD.*

Handarini, O. I., Wulandari, S. S., & Program. (2020). Pembelajaran Daring Sebagai Upaya Study From Home (SFH) Selama Pandemi Covid 19. *Jurnal Pendidikan Administrasi Perkantoran (JPAP)*, 8(3).

Masmuzidin, Z. (2018). the Current Trends of Augmented Reality in Early Childhood Education. *The International Journal of Multimedia & Its Applications*, 10(06), 47–58.

Reflianto, & Syamsuar. (2019). Pendidikan dan tantangan pembelajaran berbasis teknologi informasi di era revolusi imdustri 4.0. *Jurnal Ilmiah Teknologi Pendidikan*, 6(2), 1–13.

Risdianto, E., Yanto, M., Kristiawan, M., & Gunawan, G. (2021). *Respon Guru Pendidikan Anak Usia Dini terhadap MOOCs berbantuan Augmented Reality*. 5(2), 1487–1500.

Rosmiati, R. A. (2020). Aspects of Visual Communication Design in Animated Learning Media for Early Childhood and Kindergarten. *International Journal of Social Sciences (IJSS)*, 122–126.

Sadikin, A., & Hamidah, A. (2020). Pembelajaran Daring di Tengah Wabah Covid-19. *Biodik*, 6(2)

Saurina, N. (2016). Pengembangan Media Pembelajaran Untuk Anak Usia Dini Menggunakan Augmented Reality. *Jurnal IPTEK*, 20(1), 95.

Senduk, E. P., Sinsuw, A., & Karouw, S. (2016). M-Learning Pendidikan Karakter untuk Anak Usia Dini Berbasis Augmented Reality. *Jurnal Teknik Informatika*, 9(1), 1–5.

Sinduningrum, E., Rosalina, R., & Hilda, A. M. (2019). Pemanfaatan Teknologi Augmented Reality Untuk Media Pengenalan Huruf Alfabet Pada Anak Usia Dini. *Jurnal SOLMA*, 8(1), 142.

Sunarto, M. J. (2015). Improving Students Soft Skills using Thinking Process Profile Based on Personality Types. *International Journal of Evaluation and Research in Education (IJERE)* , pp 118–129.

Syamsuar, R. (2018). Pendidikan dan tantangan pembelajaran berbasis teknologi informasi di era revolusi industri 4.0. *E-Tech : Jurnal Ilmiah Teknologi Pendidikan.*

Yuliantina, I. (2021). Retrieved 2021, Badan Akreditasi Nasional Pendidikan Anak Usia Dini dan Pendidikan Nonformal.

Improving Assessment and Evaluation Strategies on Online Learning – Wonorahardjo, Karmina & Habiddin (eds)
© 2022 copyright the Authors, ISBN: 9781032198903

Development of Android-based interactive multimedia on interaction of living things with the environment topic for seventh grade junior high school to improve student's learning motivation

Munzil*, Y. Affriyenni & K. Kharismaliyansari
Universitas Negeri Malang, Indonesia

ABSTRACT: Attractive learning media for studying the interaction of living things with the environment is still needed because students' learning motivation with regard to this topic is still relatively low. The low motivation to learn can be due to by several factors, such as used learning materials are less attractive, teaching still involves the traditional lecture method, lack of use of interesting learning media, and a number of concepts to learn that requires students to do memorize more than to understand. To overcome these problems, it is necessary to develop an interesting learning media so that students' learning motivation on the topic can be enhanced. Hence, this research applies research and development (R&D) methods and Lee and Owens (2004) development model. The average percentage obtained overall is 93%, so it can be concluded that this learning media is valid and feasible to use.

Keywords: Android-based Interactive Multimedia, Interaction of Living Things with The Environment

1 INTRODUCTION

Science at junior high school level in Indonesia is taught in an integrated manner (Puspa et al. 2019). Because science discusses natural environment, it is hoped that the subject is preferred by most students. On the contrary, science is one of the subjects that the majority of students are less interested in. This low student interest can later affect student learning outcomes in science lessons (Karina et al. 2017). The low interest in students' learning also can be seen based on their low learning motivation. Low student motivation in science lessons is caused by a monotonous learning atmosphere and students' difficulties in understanding the material (Israil 2019). One of the science materials that was considered challenging by students is the interaction of living things with the environment, which is taught in class VII in even semesters (Rianto & Wulandari 2018).

Ghozali and Qosyim (2017) conducted a research on science material problems with regard to the interaction of living things with the environment. A science teacher at MTs Negeri Model Babat revealed that only 50% of the class VII A students completed the material. In addition, 93% of the 30 grade VII students thought that the material on the interaction of living things with the environment requires memorization and hence is difficult to understand (Ghozali & Qosyim 2017). Furthermore, students' low motivation to learn this material is also observed at SMPN 1 Dau Malang. Based on an interview with science teachers, the low motivation to learn the material is identified based on the inactivity of students that took place in the learning process. Furthermore, it can also be identified based on the number of students who did not understand the concept well based on the results of the conducted student's ability test. A questionnaire distributed to 84 seventh grade students shows that there are several reasons why students' learning motivation on the material is low, namely 58% because it consists of many concepts, 63% because they only memorize, 62% because there are too many texts in science textbooks, and 55% because they are bored.

*Corresponding Author

DOI 10.1201/9781003261346-4

The problem of low student motivation basically can be addressed by using interesting learning media (Wahyuningtyas & Rosita 2019). One of them is interactive multimedia consisting of various media elements that explain the summary of the material (Armansyah & Sulthoni 2019). This is evidenced in a research by Fauziah et al. (2016) at Tarbiyatul Aulad Cikajang Middle School. The results of the study showed that students' motivation and learning outcomes increased when interactive multimedia was used in learning. In previous research, the explanation of material in interactive multimedia was developed in various versions. There is an explanation version of the material in the form of text as well as visualization (Armansyah & Sulthoni 2019). In addition, there is also an explanation version of the material using videos equipped with multiple-choice evaluation questions (Setiawan et al. 2018).

To display interactive multimedia with the latest version, this research interactive multimedia was developed with several innovations, both in the explanation of the material and in the form of evaluation questions or quizzes. The first innovation is the addition of an audio explanation of the material. The second innovation is in the way to display the evaluation questions into games. The third and fourth innovations are using *Crossword* and *Match-Up* games to present evaluation questions or quizzes. The fifth innovation is its presentation in the form of an application that can be operated using an Android-based smartphone, wherein the aim is to make interactive multimedia displays more attractive for students. If students are interested, understanding the material will be easier (Rozi & Khomsatun 2019).

Some of these innovations are a result of method analysis carried out in this research. Such as Needs Analysis and Front-End Analysis. Front-End Analysis consists of Student Analysis, Technology Analysis, Task Analysis, Important Event Analysis, Problem Analysis, Objective Analysis, and Media Analysis. Some of the previous descriptions became the basis for research of this paper, which was carried out to produce valid and feasible learning media.

2 METHODS

The type of research conducted is *Research & Development*, which aims to produce a product (Sugiyono 2018). The development model applied is the development model by Lee and Owens (2004), because the sequence of stages is systematic and focused on developing learning multimedia. There are five stages in the Lee and Owens (2004) development model, including (1) *Assessment and Analysis*, (2) *Design*, (3) *Development*, (4) *Implementation*, and (5) *Evaluation*. The five stages are presented in Figure 1 below:

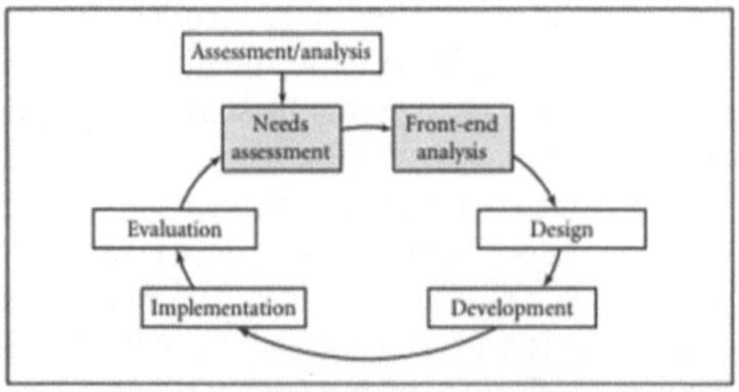

Figure 1. Lee and Owens development model procedure. Source: Lee and Owens (2004).

Due to limited research period, the study was only conducted in the development stage. Overall, the stages include *Assessment and Analysis* consisting of (1) Needs Analysis, (2) Front-End Analysis: (a) Student Analysis, (b) Technology Analysis, (c) Task Analysis, (d) Important Event Analysis, (e) Problem Analysis, (f) Objective Analysis, and (g) Media Analysis; (3) *Design*; and (4) *Development*: (a) Production, and (b) Post Production.

The validation comprises media and material validation with the subject of one lecturer in the Science Education Study Program, Faculty of Mathematics and Science, State University of Malang. The minimum criteria for the validator of the lecture are master's degree and expertise in the field of material and media developed in this research. The tests included a feasibility test and readability test. The subject of the feasibility test is one science teacher at SMPN 1 Dau Malang who is adequately educated and understands the material contained in the developed media. The subjects of the readability test are 32 students of VII grade of SMPN 1 Dau Malang who took the material on interaction of living things with the environment.

The questionnaire for validation, feasibility test, and readability test is a mixed questionnaire. Before use, the mixed questionnaire was validated by the supervisor. It is intended that the instrument used and the results obtained are declared valid. The data obtained from the mixed questionnaire are in the form of two types of data, namely quantitative data from scores and qualitative data from suggestions and comments given by the validator and the subject of the feasibility and readability test. There were two statistical scales used in the questionnaire, namely the Likert Scale and Guttman Scale.

The Likert Scale was used in the media validation questionnaire, two aspects in material validation, feasibility test, and readability test. While the Guttman Scale used to check the correctness of the concepts and the quizzes in the material validation questionnaire. The conclusion and scores on the Likert Scale are presented in Table 1, while Guttman Scale scores are presented in Table 2.

Table 1. Likert scale.

Criteria	Score
Very agree/very good/very interesting	4
Agree/good/interesting	3
Less agree/less good/less interesting	2
Disagree/not good/not interesting	1

Source: Likert (1932).

Table 2. Guttman scale.

Correctness of Concepts/Quiz	Score
True	1
False	0

Source: (Guttman (1944).

The results of the obtained assessment scores were analyzed using the following formula:

$$P = \frac{\sum X_i}{\sum X} \chi\, 100\% \quad (1)$$

Description:
P : Percentage of total score of the assessment results
$\sum X_i$: Number of test subjects' answers
$\sum X$: Maximum number of answers

The criteria for the validity and feasibility of the product can be seen from the categories in Table 3.

Table 3. Criteria for percentage of interactive multimedia assessment.

No.	Score (%)	Validity/Feasibility
1.	$\leq$ 20%	Very Invalid/Very Not Feasible
2.	21–40%	Invalid/Not Feasible
3.	41–60%	Sufficiently Valid/ Sufficiently Feasible
4.	61–80%	Valid/Feasible
5.	81–100%	Very Valid/Very Feasible

Source: Riduwan (2012).

Products with a resulted percentage of 61% can be said to be valid and feasible to use. However, products with a resulted percentage of 60% can be said to be less valid and less feasible for use. Product revisions were carried out according to the suggestion given by the validator and the subject of the feasibility and readability tests.

3 RESULT AND DISCUSSION

Interactive multimedia in previous studies was developed with material explanations using animated text and images as well as multiple-choice evaluation questions. The first example is the development of interactive multimedia with material explanations in the form of text and visualization (Armansyah & Sulthoni 2019). The second example is the development of interactive multimedia with material explanations in the form of videos but evaluation questions in the form of multiple choices (Setiawan et al. 2018). The interactive multimedia developed by Setiawan et al. (2018) can only be operated using a computer or a notebook with the appropriate specifications. To present different interactive multimedia, researchers introduced several new innovations. The latest innovations are based on the results of the analysis on the methods that have been carried out.

The developed interactive multimedia was named SIMAHL *Interactive Multimedia*. This product is operated by using an Android smartphone with minimum Android system version 2.2 Froyo (2010). There are several pages in this developed product that include the main page, profile, menu, instructions, competencies, materials, quizzes, and references. Figure 2 shows SIMAHL *Interactive Multimedia*:

Figure 2. SIMAHL *Interactive Multimedia* display.

The first innovation in SIMAHL *Interactive Multimedia* is the addition of an audio explanation of the material. The audio in this product is not only in the form of music audio but also audio explanation of the material by the narrator. It is intended that the explanation of the material is not only in the form of text but also images, animations, and audio. By explaining the material using the four media elements, SIMAHL *Interactive Multimedia* covers three forms student learnings, namely visual, audio, and kinesthetic (Fauziah et al. 2016).

The second innovation in SIMAHL *Interactive Multimedia* is in the way to display the evaluation questions into games. Previous research conducted by Setiawan et al. (2018) shows that evaluation questions are presented in the form of multiple choices. Displaying the quiz in a new way can make students more interested by designing evaluation questions or quizzes in the form of games. The game element in interactive multimedia will make learning fun (Ghozali & Qosyim 2017). Moreover, a majority of students will be interested in science learning that makes use of game media (Widianto et al. 2017).

The third and fourth innovations of SIMAHL *Interactive Multimedia* are using *Crossword* and *Match-Up* games to present evaluation questions or quizzes. Learning media using *Crossword* was proven to be able to increase students' learning motivation (Farih et al. 2012). It is possible since *Crossword* can foster students' curiosity to solve the existing questions. *Match-Up was* chosen because it can help students understand concepts well and do not just rely on rote learning (Andrea 2015).

The fifth innovation of SIMAHL *Interactive Multimedia* is its presentation the form of an application that can be operated using an Android-based smartphone. Since a majority of people now have Android smartphones, both adults and children (Mulyanto et al. 2018). Therefore, SIMAHL *Interactive Multimedia* was developed based on Android so that it can be used by junior high school students. In addition, based on research by Mulyanto et al. (2018), the presentation of Android-based applications can increase student interest in learning from 73% to 100%.

Results of validation, feasibility test, and readability test

Based on the validation, feasibility test, and readability test that have been carried out, the results are presented in Table 4.

Table 4. Result of validation, feasibility test, and readability test.

No.	Aspects	Analysis Result (%)	Assessment Criteria
1.	Media Validation	91	Very Valid
2.	Material Validation	95	Very Valid
3.	Feasibility Test	96	Very Feasible
4.	Readability Test	91	Very Feasible
Average		93	Very Valid & Feasible

Based on Table 4, it can be seen that the results of media validation with three assessment aspects yielded 91% results, while the results of material validation with four assessment aspects show 95% results. From these results, it is known that SIMAHL *Interactive Multimedia is* valid for use. Media and material validation was carried out by a lecturer at the Science Education Study Program, Faculty of Mathematics and Natural Sciences, State University of Malang. In addition, the results of the feasibility test with five assessment aspects got a result of 86% and the result of the readability test with 17 assessment items got a result of 91%. From the two percentage results, it is known that SIMAHL *Interactive Multimedia is* feasible to use. The feasibility test was carried out by one science teacher at SMPN 1 Dau, while the readability test was carried out by 32 students at SMPN 1 Dau who had taken the material on the interaction of living things with the environment. From the four results, an average of 93% was obtained so that SIMAHL *Interactive Multimedia* can be declared valid and feasible to use.

In general, it can be concluded that interactive multimedia is declared valid and suitable for use in learning. This is also proven by research by Setiawan et al. (2018) and Armansyah and Sulthoni (2019). From the research of Setiawan et al..(2018), interactive multimedia in the form of 3D digital books developed obtained validity of 92.33%, practicality of 88.01%, attractiveness of 91.85%, and effectiveness of 80.66%. Meanwhile, from the research of Armansyah and Sulthoni (2019), interactive multimedia on visualization based on animation obtained presented material validity of 70%, media validity of 85%, and audience assessment of 81.7%. From the two studies, the researchers also concluded that the interactive multimedia developed was valid and feasible to use.

4 CONCLUSION

Based on the results above, it is known that the SIMAHL *Interactive Multimedia* got an average result of 93%. Based on these results it can be concluded that SIMAHL *Interactive Multimedia* is valid and feasible to use for increasing students' learning motivation. In addition to SIMAHL *Interactive Multimedia* display, interestingly, evaluation questions or quizzes in the form *Crossword* and *Match-Up* can also make students interested in learning material on the interaction of living things with the environment.

Based on this research, there are several suggestions that can be applied for further research. The first suggestion is that the learning media should be developed not only based on Android, but can be operated using any smartphone. The second suggestion, to increase the attractiveness of learning media, researchers can develop material explanations and evaluation questions in the form of games that are preferred by many students. The third suggestion is that evaluation questions must be developed so that they can reach students' higher-order thinking processes. The fourth suggestion is that research should be applied to the evaluation stage.

REFERENCES

Andrea, R. (2015). Teknik Pengacakan Posisi Objek Permainan Match-Up "Find Me! – Bumi Etam." *Seminar Nasional Teknologi Informasi Dan Multimedia*, 3(1), 1–4.

Armansyah, F., & Sulthoni, S. (2019). Multimedia Interaktif sebagai Media Visualisasi Dasar-Dasar Animasi. *Jurnal Kajian Teknologi Pendidikan*, 2(3), 224–229.

Farih, Mansur, Marimin, & Ismiyati. (2012). Penerapan Teka-Teki Silang pada Stenografi dengan Model Stad untuk Meningkatkan Motivasi. *Economic Education Analysis Journal*, 1(1), 1–6.

Fauziah, F., Setiawan, D., & Rahadian, D. (2016). Efektivitas Penggunaan Multimedia Interaktif terhadap Motivasi dan Hasil Belajar Siswa di SMP Pada Mata Pelajaran IPS. *Jurnal Teknologi Pendidikan Dan Pembelajaran*, 1(1), 26–37.

Ghozali, M. S., & Qosyim, A. (2017). Efektifitas Permainan Teka-Teki Silang Tiga Dimensi (TTS3D) sebagai Media Pembelajaran pada Materi Interaksi Makhluk Hidup. *E-Jurnal Pensa*, 5(1), 243–247.

Guttman, L. (1944). A Basis for Scaling Qualitative Data. *American Sociological Review*, 9(2), 139.

Israil, I. (2019). Implementasi Model Pembelajaran Cooperative Learning Tipe STAD untuk Meningkatkan Motivasi Belajar Siswa dalam Pembelajaran IPA di SMP Negeri 1 Kayangan. *Jurnal Kependidikan*, 5(2), 117–123. https://doi.org/10.33394/jk.v5i2.1807

Karina, R. M., Syafrina, A., & Habibah, S. (2017). *Hubungan antara Minat Belajar dengan Hasil Belajar Siswa dalam Mata Pelajaran IPA pada Kelas V SD Negeri Garot Geuceu Aceh Besar*. 2(1), 61–77.

Lee, W. W., & Owens, D. L. (2004). *Multimedia-Based instructional Design*. San Francisco: Preiffer.

Likert, R. (1932). *A Technique for The measurement of Attitudes*. New York: New York University.

Mulyanto, A., Apriyadi, & Prasetyawan, P. (2018). Rancang Bangun Game Edukasi "Matching Aksara Lampung" Berbasis Smartphone Android. *Journal of Computer Engineering System and Science*, 3(1), 36–44. https://doi.org/10.24114/cess.v3i1.8225

Puspa, M. A., Gobel, C. Y., & Djafar, A. (2019). Aplikasi Pembelajaran IPA untuk Kelas VIII di Sekolah SMP Negeri 1 Pulubala Kabupaten Gorontalo Berbasis Android. *Jurnal Informatika Upgris*, 5(1), 1–8.

Rianto, B. W., & Wulandari, T. S. H. (2018). *Pengembangan Lembar Kerja Siswa (LKS) Berbasis Mind Mapping untuk Meningkatkan Penguasaan Konsep Belajar pada Materi Interaksi Makhluk Hidup dan Lingkungannya*. 15(1), 459–465.

Riduwan. (2012). *Pengantar Statistika Sosial*. Bandung: Alfabeta.

Rozi, F., & Khomsatun, K. (2019). Rancang Bangun Game Edukasi Pengenalan Warna untuk Pendidikan Anak Usia Dini Menggunakan Adobe Flash Berbasis Android. *JIPI (Jurnal Ilmiah Penelitian Dan Pembelajaran Informatika)*, 4(1), 12–18. https://doi.org/10.29100/jipi.v4i1.781

Setiawan, D. A., Wahjoedi, & Towaf, S. M. (2018). Multimedia Interaktif Buku Digital 3D pada Materi IPS Kelas IV Sekolah Dasar. *Jurnal Pendidikan*, 3(9), 1133–1141.

Sugiyono. (2018). *Metode Penelitian Pendidikan: Pendekatan Kuantitatif, Kualitatif, dan R&D*. Bandung: Alfabeta.

Wahyuningtyas, N., & Rosita, F. A. D. (2019). Pengembangan Multimedia Interaktif Berbasis Android pada Materi Kehidupan Sosial Masyarakat Indonesia. *Jurnal Sejarah, Budaya Dan Pengajarannya*, 13(1), 34–41. https://doi.org/10.17977/um020v13i12019p034

Widianto, T. R., Raharjo, & Rosdiana, L. (2017). Pengembangan Permainan Kartu Uipa Untuk Meningkatkan Hasil Belajar Siswa Pada Materi Interaksi Antar Makhluk Hidup. *Pensa: Jurnal Pendidikan Sains*, 5(1), 21–26.

Improving Assessment and Evaluation Strategies on Online Learning – Wonorahardjo, Karmina & Habiddin (eds)

U-learning: Rethinking assessment for 21st-century learners

M.F. Yahaya
Universiti Teknologi Mara Perlis Branch, Malaysia

ABSTRACT: The digital transformation resulting because of continuous developing of new technologies and the surge of Covid-19 pandemic is evidence of how Volatile, Uncertain, Complex and Ambiguous (VUCA) the world is. To ensure that learning is not disrupted during the pandemic when physical classrooms are closed, educators have to manipulate the available resources and experiment with existing technologies, so that learning and teaching can occur ubiquitously. Nevertheless, teaching and learning are far from complete if they are not accompanied by assessment. Due to this belief, this paper focuses on assessment as part of ubiquitous learning to explore whether the existing assessment prepares the 21st-century generation for their future. Specifically, the objectives of the paper are twofolds: first, it describes the current trend in publication on assessment and u-learning, and second, suggests future assessment for 21st-century learners. The paper begins by discussing the challenges in the assessment of 21st-century learners. This is followed by discussions on the publication trend of research on assessment and the future of assessment for 21st-century learners. Pedagogical and research implications are also discussed.

Keywords: 21st-century learners, VUCA, U-learning assessment

1 INTRODUCTION

Towards the end of 1999, the world was warned about the millennium bugs. The computerised coding of time and date had been named with the source of the problems affecting digital information (Britannica 2021). Hence, measures were taken by relevant parties to ensure the impact of Y2K (the Year 2000) was managed effectively. The date 1st January 2000 was not just to remember the successful collective efforts of handling the millennium bugs, but also marked the beginning of the 21st century.

Meanwhile, the learners were called the 21st-century learners or Generation Z. Dimock (2019), however, reclassified Generation Z as those born between 1997 and 2012. This year, these learners should be between the ages 9 and 24, and many of them should either be in schools or colleges.

Being post-millennial generation, they would probably encounter more bugs and challenges than other generations before them. What would be more memorable with this generation is that they are welcomed by the internet or web 1.0, which existed before them. This generation and the coming generations have grown to learn, love, and experience whatever elements the internet can offer. The continuous and rapid development of the internet has triggered endless digital transformations in all aspects of life, including economy, society and education (Tolboom 2016).

This has accelerated the application of the VUCA world concept, an acronym for Volatility, Uncertainty, Complexity and Ambiguity. This VUCA world concept, as it is known, provides a fresh breath of explanation and justifications and has been applied to reflect the dynamics of the

DOI 10.1201/9781003261346-5

changes. VUCA has also been used to explain the impact of the Covid-19 pandemic (Nangia & Mohsin 2020). Similarly, the VUCA concept has also been used in describing the changes that have taken place in education. Due to these attributes, the VUCA concept will also be manipulated to explain the current subject matter.

The Covid-19 pandemic has changed the whole education landscape. Both the educators and students have been forced to work from a distance and explore new avenues of teaching and learning that they might have never experienced before. At the moment of testing time, the traditional physical classroom has been seen as dangerous and impractical. Instead, a flexible learning concept known as ubiquitous learning or u-learning, where learning can be done anywhere and anytime, would be a more practical and sensible approach (Hwang et al., 2008). However, learning is not complete if it is not complemented by assessment (Deeley et al. 2019; Taras 2002). The change in the learning concept has become a concern due to the need to match assessment and the U-learning of 21st-century learners. Hence, the aim of this paper is to explore whether the existing assessment prepares 21st-century learners for their future. Specifically, the objectives of this paper are twofolds: first, describe the current publication trend on assessment in u-learning and second, predict and discuss future challenges in assessment.

2 21ST-CENTURY LEARNERS AND TECHNOLOGY

Amid this VUCA world, 21st-century learners have endured and experienced rapid changes creating a robust generation to technological and digital transformation (Onyema & Daniil 2017; Rof et al. 2020). The continued development in digital technology has taken a new acceleration peddle with the Industrial Revolution 4.0. The development has been so rapid that changes in technology happen within hours. Apparently, this generation of 21st-century learners has to acclimatise to these changes in technology. In many instances, their subtle approaches to technologies have put them ahead of their predecessors. Hence, it is expected that these 21st-century learners are also known as "digital natives".

As digital natives, they are expected to be highly independent and savvy with technologies (Hirschman & Wood 2018). The issue is whether they are all "digital natives". This notion of being digital natives was probably because they came to existence just when digital technologies started to gain popularity. The rapid development of technologies has forced them to embrace and adopt digital skills. However, according to Neumann (2016), not all 21st-century learners are digital natives. Some may have difficulties coping with technological demands. Hence, it is the role of education to nurture and channel these learners into becoming what they are expected to become. But are we moving in the right direction?

2.1 *The development of education*

The purpose of education is to prepare students for the future and beyond (Driscoll n.d). In doing so, the curriculum, syllabus, teaching and assessment should foster and outline the students' future. This is upon consideration that the existing educations have probably been designed, prepared and used by either the baby boomers (born between 1946 and 1964) or the generation X (born between 1965 and 1980). However, does the existing education prepare them for what they will embark on in the future?

The spillover effects of the accelerated digital transformation have also descended on education. These changes in education are imminent as newer technologies are both adapted and adopted, leading to ubiquitous learning or u-learning. The concept of ubiquitous learning has been used to refer to the phenomena in which learning can happen anywhere and at any time (Hwang et al. 2008). In those instances, learning can be conducted using mobile or static devices (depending on the availability of the devices to the students) and connected to a network (Hwang et al. 2008). The role of the teachers has also changed from being the sole providers of knowledge (Sjöberg & Lilja 2020) to being designers of lessons due to the unlimited resource available digitally.

However, the success of ubiquitous learning depends on the environmental awareness of both the teachers and the students. This concept of context-aware ubiquitous learning has been the determining factor of a classrooms' success. Before the pandemic, at one point or the other, some opponents were against the approach of conducting classes ubiquitously. However, the Covid-19 pandemic forced them to change their stand into teaching online ubiquitously (Lin et al. 2020). Hence, it was safe for anyone to say that 80% of all classes have been conducted ubiquitously at all levels.

The sudden changes in the situation can be attributed to the volatile, uncertain, complex and ambiguous world phenomena. Nevertheless, the call for a continued education rather than an abruptly halted education has driven the exploration of other learning options. In doing so, the learners and teachers are required to acquire context-aware knowledge and experiment with appropriate approaches and technologies to suit the context.

These elements, which reflect the criteria of ubiquitous learning (Hwang et al., 2008), mean the success of education does not depend solely on the tools or device that are used; instead, it depends on the context where the learning takes place as well as the tools or devices that suit the context (Aljawarneh 2020). This also means that the "one shoe fits all" approach cannot be applied. The issue is whether the existing students and educators can match the context and the technology? What impact would it have on assessments?

3 VUCA, COVID-19 AND THEIR CHALLENGES TO EDUCATION

VUCA phenomena is a relatively new concept, and it was first used in the late 1980s. The idea, nevertheless, has gained its momentum ever since the Covid-19 outbreak when VUCA had been applied to rationalise the state of the pandemic (Murugan et al. 2020; Nangia & Mohsin, 2020a; Noda 2020;. The V for the volatility of the pandemic can be observed from the dynamics of the changes that revolve around the pandemic. Meanwhile, the U in uncertainty refers to the inability to forecast the chain of events due to the pandemic. The C in complexity reflects the enormity of the impact of the pandemic. Finally, the A in ambiguity refers to the lack of ultimate standard procedure to deal with the pandemic.

In addition to being used to explain the impact of the Covid-19 pandemic, VUCA can also be applied to explain the impact of the pandemic on education. Education throughout the world has been dramatically affected by the surge in the Covid-19 cases. As a precautionary measure to control the spread of the virus, schools in many parts of the world have been closed at one time or the other. Contact between students and teachers have been limited to distant online meeting or cloud meeting.

3.1 *Challenges in assessment*

Learning is not complete without assessment. In fact, assessment has been used as an integral segment of teaching and learning. Among the purposes of assessment is to measure how successful the process of teaching and learning is. In some learning contexts, the assessment itself is learning. According to Khan (2012), assessment can improve learning. This happens when the lesson begins with assessment (diagnostic), assess the students while having the lesson (formative) and ends the class with assessment (summative). In this situation, assessments are used throughout the lesson. Most importantly, assessment is used to measure the extent to which the learning outcome is achieved (Malaysian Qualification Agency 2011). This means that assessments should reflect what they intended the students to do upon completing their studies.

Assessing the students is always challenging regardless of any circumstances. In discussing the challenges of assessment, it is essential to view assessment or evaluation from the commonly used principles of assessment, namely validity, flexibility, reliability and fairness. These principles have been used in developing and evaluating any assessment. An assessment has to be valid, flexible, reliable and fair. Any assessment that does not meet the principle would have to go through some levels of revisions to ensure that it is within the acceptable standards of its principles.

An assessment is valid if it successfully measures what it intends to measure (Teglasi et al. 2012). In the context of university courses, the assessment should match the learning outcome that has been set. If it differs from the learning outcome, it is, therefore, assumed to be not valid. The learning outcome reflects what the educators wanted the students to do and what they needed to know or master. In other words, assessment should measure their readiness for their future. But the issue is whether the assessment really measures their readiness for their future or others?

Meanwhile, an assessment is reliable if the approach to the assessment and the measurement of the values are consistent (Sullivan 2011). The key to reliable assessment is a consistent result regardless of who the assessors are. This perhaps can be obtained from the standards set in all aspects of the assessment, including the administration, the structure of the test, and the rubrics used. However, in a VUCA context like the pandemic, when educators have to alter their approach to suit the situation, would the assessment still be reliable?

Concerning flexibility, an assessment is flexible when it considers the individual needs of the people involved in the assessment. This is a difficult feat to handle, especially when you have a diverse group of individuals sitting for the test. Furthermore, in a state of volatility, uncertainty, complexity and ambiguity, flexibility may be problematic.

Finally, an assessment is fair if everyone has equal opportunities to excel. In a normal situation, everyone present for the assessment will be tested in the same context. This would create certain levels of fairness. However, the VUCA context may affect the degree of fairness. How can it be considered fair if the participants are at different locations, using various devices and at different levels of internet connections?

Eventually, ensuring that each assessment is within the scope of the principle of assessment may not be simple as there would be several variables that may affect the outcome of the principle of assessment. The impact of VUCA on the development and administering of the assessments are among the elements that can affect the principles, the integrity and the success of the assessment. Whatever the challenges or issues, educators have to find ways to address these challenges and issues.

4 PUBLICATION TREND OF U-LEARNING ASSESSMENT FOR 21ST CENTURY LEARNERS

The purpose of looking into trends is not only meant to identify the situations but also to predict or forecast their future. In this context, analysis of the publication trend of U-learning assessment for 21st-century learners would enable the researcher to gauge the current situation on the U-learning assessment for 21st-century learners.

In achieving this feat, publication data were extracted from two repositories: Scopus and Web of Science, on 1st July 2021. The use of the Prisma flow diagram was deemed to be suitable to describe the process involved in identifying and selecting the articles.

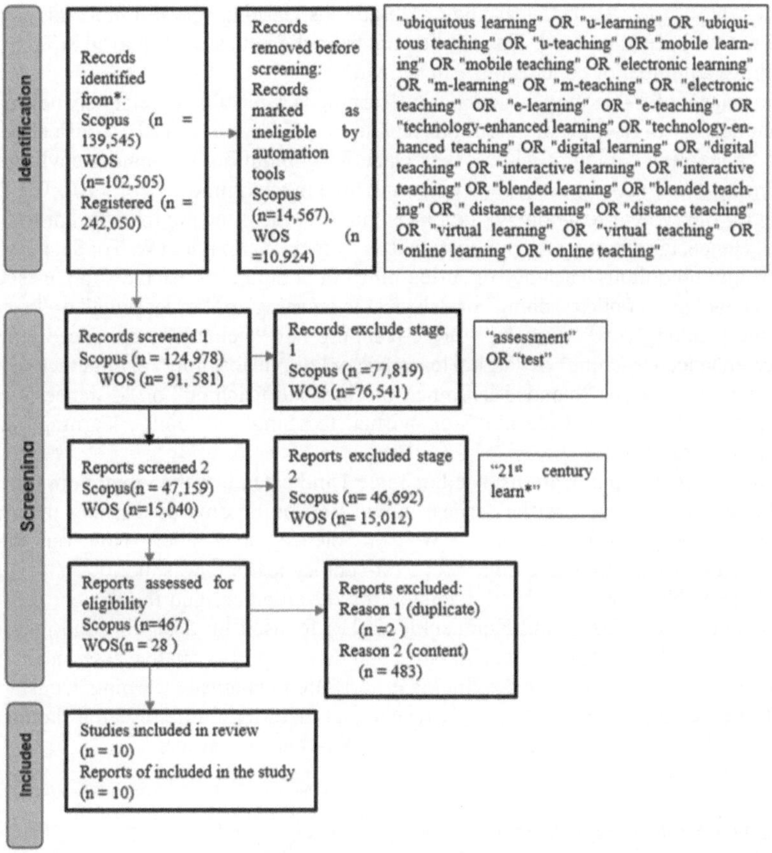

Figure 1. Prisma flow diagram on the procedures used to select articles.

The process, as shown in Figure 1, enabled relevant documents to be selected. Based on the Prisma flow diagram, only 10 documents were found to suit the criteria set. The list of documents can be found in Table 1 below.

Table 1. Publication on U-learning assessment for 21st-century learners.

Authors	Year	Issues
Khlaisang, J., Koraneekij, P.	2019	Measurement for current use in various online learning systems
Spector et al.	2016	Technologies supporting formative assessment and informative feedback
Pospisilova, L.	2018	ePortfolio in language learning with the pedagogical principles
McKenna, H.P., Chauncey, S.A.	2015	Consensual assessment technique (CAT) to assess creativity, innovativeness, and value of the product
Redecker et al.	2012	Testing paradigm
McKenna et al.	2013	Adaptation of the consensual assessment technique
Nikou, S.A., Economides, A.A.	2016	Mobile-based Micro-Learning and Assessment (MBmLA) homework activities
Khlaisang J., Koraneekij P.	2018	Open Online Assessment Management System and standardised measurement
Allen, J.M., Wright, S., Innes, M.	2014	Higher education online learning and assessment
Bacca-Acosta, J., Avila-Garzon, C.	2021	Student engagement with mobile-based formative assessment

Based on Table 1, the area of U-learning or Ubiquitous learning was still in its infancy. A lot of issues have not yet been addressed. This can be seen from the issues, as displayed in Table 1. Based on the findings, a number of conclusions can be made.

First, the concept of U-learning or Ubiquitous learning is a general but relatively new concept of learning that deals with the idea of ubiquitous technology. Yahaya et al. (2010) defined ubiquitous learning as "a learning paradigm which takes place in a ubiquitous computing environment that enables learning the right thing at the right place and time in the right way" (pg 120). The definition did not specify the type of technology, the place and the time allowing the concept to be further generalised. Hence, in extracting the publication data from the Scopus and Web of Science database, the word strings "ubiquitous learning" or "u-learning" or "ubiquitous teaching" or "u-teaching" or "mobile learning" or "mobile teaching" or "electronic learning" or "m-learning" or "m-teaching" or "electronic teaching" or "e-learning" or "e-teaching" or "technology-enhanced learning" or "technology-enhanced teaching" or "digital learning" or "digital teaching" or "interactive learning" or "interactive teaching" or "blended learning" or "blended teaching" or "distance learning" or "distance teaching" or "virtual learning" or "virtual teaching" or "online learning" or "online teaching" were employed.

Second, the number of publications listed in Table 1 indicated that the research in this area was limited. This can be expected as the earliest publication on ubiquitous learning in three major repositories, Google Scholar, Scopus and Web of Science, dated 2002. However, the earliest publication data on U-learning assessment for 21st-century learners was in 2012.

Third, all the publications in Table 1 focus on the development and the design of the assessment. For example, Khlaisang and Koraneekij (2019) focused on designing the measurement, Spector et al. (2016) focused on the technologies that can support formative assessment and feedback, Pospisilova (2018) focused on developing e-portfolio in language learning. Considering that assessment in U-learning classes has not yet reached its maturity, future research should consider developing and designing assessments to suit these 21st-century learners.

5 THE FUTURE OF ASSESSMENT

The close relationship between assessment and teaching or learning means that assessment should match teaching and learning. Specifically, assessment should match the learning outcome, the syllabus, the content and the students. The inability of the students to perform in the assessment indicated that the students had not attained the standard set by the learning outcome. However, the ultimate goal of education is not about learning outcomes; instead, it should prepare the learners for the future. How do you design an assessment for the future? Shouldn't it be reflecting what and how the students should be performing in future?

Thus in reflecting the students' performance in the future, the assessment for the future should not be about memorising or recalling the correct answers. Upon completing their studies, they will have to obtain, organise and design solutions based on available resources. At that time, they will be allowed and expected to use the information beyond what they have in their memories. Therefore, the future of assessment should enable students to be resourceful in obtaining accurate information from the available resources and applied in the current context. In fact, at the moment, when classes and assessments are conducted online, our students are actually manipulating the system by assessing materials from multiple resources. Therefore, it is only fitting that we create a system of assessment that allows students to access the resources instead of testing their memory.

6 CONCLUSION

In light of the VUCA phenomena, the unpredictability of the situation and the possibility of radical change can defy one's logical mind. This was evident during the first year of the pandemic as the schools were closed, and classes were conducted via a new ubiquitous platform. Although it was claimed that some teachers and students were able to cope with the new norms, others were struggling. However, the impact of the pandemic has subsided as both the educators and students have grown to cushion its effects and have become familiar with the situation. What they undergo, at this moment, may have been developed from over a year of experimentation.

With the assumption that the pandemic will resolve, should the practice of teaching, learning and assessment be reverted to the time before the pandemic? Or shouldn't we progress from here? Shouldn't we just improve, modify and change whatever is necessary.

In dealing with the situations, education of the future should be resilient to the test of time, not withered when encountering more challenging conditions. This means the whole concept of education (teaching, learning and assessment) has to have flexible and agile characteristics.

REFERENCES

Aljawarneh, S.A. 2020. Reviewing and exploring innovative ubiquitous learning tools in higher education. *Journal of Computing in Higher Education 32*, 57–73. doi:10.1007/s12528-019-09207-0

Allen, J. M., Wright, S., & Innes, M. 2014. Pre-service Visual Art Teachers' Perceptions of Assessment in Online Learning. *Australian Journal of Teacher Education, 39*(9), 1–17.

Bacca-Acosta, J., & Avila-Garzon, C. 2021. Student engagement with mobile-based assessment systems: A survival analysis. *Journal of Computer Assisted Learning*, 37(1), 158–171.

Britannica, T. Editors of Encyclopaedia. 2021, 21st April. Y2K bug. Encyclopedia Britannica.

Deeley, S.J., Fischbacher-Smith, M., Karadzhov, D., Koristashevskaya, E., 2019. Exploring the 'wicked' problem of student dissatisfaction with assessment and feedback in higher education. *Higher Education Pedagogies 4*, 385–405.. doi:10.1080/23752696.2019.1644659

Dimock, M. 2019, 17th January. Defining generations: Where millennials end and generation z begins. *Pew Research Center*. Retreived 16th May, 2021.

Driscoll, M. n.d. Education in the 21^{st}century. Retrieved 18th July, 2021.

Hirschman, K., & Wood, B. 2018. 21st century learners: Changing conceptions of knowledge, learning and the child. *New Zealand Annual Review of Education, 23*, 20–35.

Hwang, G. J., Tsai, C. C., & Yang, S. J. 2008. Criteria, strategies and research issues of context-aware ubiquitous learning. Journal of Educational Technology & Society, 11(2), 81–91.

Khan, B. 2012. Relationship between assessment and students' learning. *International Journal of Social Sciences and Education, 2*(1), 576

Khlaisang, J., & Koraneekij, P. 2018. Development and validation of the online Assessment Management System for 21st century learners. In Edulearn 18. 10th International Conference on Education and New Learning Technology:(Palma, 2nd-4th of July, 2018). Conference proceedings (pp. 3852–3860). IATED Academy.

Khlaisang, J., & Koraneekij, P. 2019. Open Online Assessment Management System Platform and Instrument to Enhance the Information, Media, and ICT Literacy Skills of 21 st Century Learners. *International Journal of Emerging Technologies in Learning, 14*(7).

Lin, C., Barrett, N. E., & Liu, G. 2020. English outside the academic sphere: A mobile-based context-aware comparison study on collaborative and individual learning. *Journal of Computer Assisted Learning, 37*(3), 657–671. doi:10.1111/jcal.12514

Malaysian Qualification Agency. 2011. *Guidelines to good practices: Curriculum design and delivery*. Malaysian Qualification Agency.

McKenna, H. P., & Chauncey, S. A. 2015. Exploring a creativity assessment technique for use in 21st century learning, library, & instructional collaborations. In *8th International Conference of Education, Research and Innovation (iCERi2015)* (pp. 1713–1717).

McKenna, H. P., Arnone, M. P., Kaarst-Brown, M. L., McKnight, L. W., & Chauncey, S. A. 2013. Application of the consensual assessment technique in 21st century technology-pervasive learning environments. In *Proceedings of the 6th international conference of education, research and innovation (iCERi2013)* (pp. 6410–6419).

Page, M.J., Mckenzie, J.E., Bossuyt, P.M., Boutron, I., Hoffmann, T.C., Mulrow, C.D., Shamseer, L., Tetzlaff, J.M., Akl, E.A., Brennan, S.E., Chou, R., Glanville, J., Grimshaw, J.M., Hróbjartsson, A., Lalu, M.M., Li, T., Loder, E.W., Mayo-Wilson, E., Mcdonald, S., Mcguinness, L.A., Stewart, L.A., Thomas, J., Tricco, A.C., Welch, V.A., Whiting, P., Moher, D. 2021. The PRISMA 2020 statement: an updated guideline for reporting systematic reviews. BMJ n71. doi:10.1136/bmj.n71

Murugan, S., Rajavel, S., Aggarwal, A. K., & Singh, A. 2020. Volatility, Uncertainty, Complexity and Ambiguity (VUCA) in Context of the COVID-19 Pandemic: Challenges and Way Forward. *International Journal of Health Systems and Implementation Research, 4*(2), 10–16.

Nangia, M., & Mohsin, F. 2020a. Identifying VUCA factors in a pandemic era–a frame-work focused on Indian IT industry. *Journal of Critical Reviews, 7*(7), 931–936.

Nangia, M., & Mohsin, F. 2020b. Revisiting talent management practices in a pandemic driven VUCA environment-a qualitative investigation in the Indian IT industry. *Journal of Critical Reviews, 7*(7), 937–942.

Neumann, Crystal. 2016. Teaching digital natives: promoting information literacy and addressing instructional challenges. *Reading Improvement, 53*(3), 101.

Nikou, S. A., & Economides, A. A. 2016. The impact of paper-based, computer-based and mobile-based self-assessment on students' science motivation and achievement. *Computers in Human Behavior*, 55, 1241–1248.

Noda, Y. 2020. Socioeconomical transformation and mental health impact by the COVID-19's ultimate VUCA era: Toward the New Normal, the New Japan, and the New World. *Asian Journal of Psychiatry, 54,* 102262–102262

Onyema, O. G., & Daniil, P. 2017. Educating the 21st century learners: are educators using appropriate learning models for honing skills in the mobile age?. *Journal of Entrepreneurship Education, 20*(2), 1–15.

Pospisilova, L. 2018. Individualised and autonomous learning and assessment with ePortfolio use. In Edulearn 18. *10th International Conference on Education and New Learning Technology*:(Palma, 2nd-4th of July, 2018). Conference proceedings (pp. 3026–3032). IATED Academy.

Redecker, C., Punie, Y., & Ferrari, A. (2012, September). eAssessment for 21 st Century Learning and Skills. In *European Conference on Technology Enhanced Learning* (pp. 292–305). Springer, Berlin, Heidelberg.

Rof, A., Bikfalvi, A., Marquès, P., 2020. Digital Transformation for Business Model Innovation in Higher Education: Overcoming the Tensions. *Sustainability 12*, 4980. doi:10.3390/su12124980

Sjöberg, J., & Lilja, P. 2020. University Teachers' Ambivalence about the Digital Transformation of Higher Education. *International Journal of Learning, Teaching and Educational Research, 18*(13).

Spector, J. M., Ifenthaler, D., Sampson, D., Yang, J. L., Mukama, E., Warusavitarana, A., ... & Gibson, D. C. 2016. Technology enhanced formative assessment for 21st century learning. 4th *International EDUsummIT*

Sullivan, G.M. 2011. A primer on the validity of assessment instruments. *Journal of Graduate Medical Education, 3*, 119–120. doi:10.4300/jgme-d-11-00075.1

Taras, M. 2002. Using assessment for learning and learning from assessment. *Assessment & Evaluation in Higher Education*, 27(6), 501–510.

Teglasi, H., Nebbergall, A. J., & Newman, D. 2012. Construct validity and case validity in assessment. *Psychological Assessment, 24*(2), 464–475. https://doi.org/10.1037/a0026012

Tolboom, I. 2016. *The impact of digital transformation*. Master Thesis Report. Delft University of Technology.

Improving Assessment and Evaluation Strategies on Online Learning – Wonorahardjo, Karmina & Habiddin (eds)

Development website of planning, writing, and publication of scientific articles based on Classroom Action Research (CAR) to increase teacher's pedagogical competence

H.R. Widarti*, Habiddin, N.C.E. Setiawan, Parlan, A.B. Syafruddin & A. Ardyansyah
Universitas Negeri Malang, Indonesia

D.A. Rokhim
Chemisty, SMAN 3 Sidoarjo, Indonesia

ABSTRACT: Teachers can develop their professional skills through Continuous Professional Development (PKB) activities. One form of teacher professional development is the publication of scientific papers. Teachers still have not conducted Classroom Action Research (CAR) in a sustainable and consistent manner, including the teacher of SMAN 3 Sidoarjo, who has low ability in writing scientific papers. This study aims to produce a CAR-based educational website for scientific articles. The type of research used is research and development (R&D) using the ADDIE model with descriptive analysis techniques. The results show that the use of planning websites and writing scientific articles can overcome obstacles such, as difficulty finding ideas and lack of writing skills. This website has several excellent features, namely planning, writing, publishing articles, materials, and chat features. Thus, it can be seen that the CAR-based scientific article training website can improve the pedagogic competence of teachers.

Keywords: Scientific Articles, CAR, Website, Teacher Pedagogics

1 INTRODUCTION

The teacher is an important and collaborative actor along with parents and other important elements (Schuster et al. 2021). Without the active involvement of teachers, education will be meaningless and empty of material, essence, and substance. The teacher position is a very important professional position in achieving the vision of education (Darmadi 2015; Muchson et al. 2021). The demand for teacher professional development is increasingly felt after the issuance of Law Number 14 of 2015 concerning teachers and lecturers (Aisyah & Mahanani 2017). On the other hand, as educators, teachers are also professionals who are constantly required to develop their professional skills. Several teacher professional development activities can be carried out to achieve educational success (Basturk et al. 2020)

Professional teaching is a job that a teacher does that requires certain skills (Christoforidou & Kyriakides 2021). Teachers can develop their profession through Continuous Professional Development (CPD) activities. CPD is the development of teacher competencies that is carried out as needed, gradually, continuously to improve professionalism (Riyadi, 2016). The CPD strategy includes three aspects, namely: 1) self-development strategies through mentoring and training, 2) scientific publication strategies are divided into three groups, and 3) innovative work strategies in the form of finding appropriate technologies, discovering or developing works of art, and

*Corresponding Author

DOI 10.1201/9781003261346-6

making or modifying learning tools (Fatkuroji 2019; Fitriyadi 2013). One of the activities of CPD is Classroom Action Research (CAR).

CAR is a reflective study by action actors, aimed at increasing the rational stability of their actions (Nurgiansah et al. 2021). The results of CAR can be used as scientific articles by the teacher. Teachers experience difficulties in writing scientific articles and lack of information about the journal to be addressed (Gunawan et al. 2018). Teachers actually already have written material to make scientific articles, namely from the results of CAR conducted by teachers (Sagala et al. 2019). Therefore, it is necessary to provide assistance in writing scientific articles for teachers so that the existing materials can be followed up to be written in the format of scientific articles.

Teachers currently have difficulties in writing scientific research articles that meet the requirements for publishing scientific papers in a scientific journal, so that the intellectual work produced by teachers in scientific journals is still very limited (Sumartini et al. 2019). Teachers have duties/functions only in teaching so they do not have the habit of researching and writing scientific articles as is done by lecturers (Junaid et al. 2020). Teacher assistance needs to be made more accessible using today's technology, one of which is a website.

Along with the times, learning websites are expected to be able to bring the learning process into a more dynamic digital world (Kuncoro 2014). Websites can help teachers effectively and efficiently understand scientific articles to support the use of technology in schools. The use of the web is expected to foster teacher literacy about scientific articles. Therefore, it is necessary to have a website that can assist teachers in designing, writing, and publishing scientific articles based on CAR.

2 METHODS

The ADDIE model is structured programmatically with systematic sequences of activities in an effort to solve problems related to training needs for making scientific articles based on CAR (Kurnia et al. 2019). In accordance with the chosen model, the first stage is analyzing needs and problems in the form of relevant material analysis, front-end analysis, teacher analysis, task analysis, concept analysis, and determining instructional objectives. In the ADDIE stage, an analysis of the basic problems is carried out based on the CAR used, analysis of guidebooks, planning, writing, and publication of scientific articles; (2) at the design stage, several activities are carried out including the selection of the media used, display format or storyboard design, and initial design related to the interactive website media components; (3) the stage of development, preparation and writing of material in a guidebook is tailored as per the needs of teachers in making scientific articles. In the ADDIE stage, five aspects were explored in the expert validation stage, including media format, language, media content, graphics, and aspects of media presence, as well as development trials carried out with readability tests.

Stage four is the implementation of product development in the form of training materials on planning, writing, and publishing scientific articles. (5) Evaluation stage is carried out formatively at the product development stage in accordance with the model used. The validator will validate the website media that has been created as a means of training scientific articles for teachers.

The data obtained will be in the form of qualitative and quantitative data. Explanatory items from qualitative data are comments, suggestions, or criticisms from validators. While the explanatory items from quantitative data are numbers obtained from filling out the validation instrument sheet for product development using a Likert scale (5,4,3,2,1). The quantitative item analysis technique used is the average calculation technique as follows (Alan & Atalay Kabasakal 2020):

$$\tilde{X} = \frac{\Sigma x}{n}$$

Under:
X: average value
x : number of validator answers/assessment of test subjects
n: number of validators

To find the conclusion from the results of the average calculation above, the range of validation criteria is used. The range of validation criteria for the complete calculation of the results can be observed in Table 1.

Table 1. Criteria for validation results

Value	Criteria
–20%	Very Poor
20–40%	Poor
40–60%	Average
60–80%	Good
80–100%	Very Good

Source: Alan & Ataly Kabasakal (2020)

3 RESULT AND DISCUSSION

3.1 *Developed model*

The website for planning, writing, and publishing scientific articles based on CAR developed by the researcher is named Smart-Teacher: A training website for making scientific articles. This article training website was developed as a follow-up to the results of CAR that has been carried out by the teacher. This website was developed by displaying several sub-materials that support understanding related to making scientific articles (Susilowati 2018). The website also displays a page about some information about the journal according to the chosen topic. Scientific article templates have been provided in selected journals.

Websites were developed because they are easily accessible and flexible. The website can be operated using a smartphone or laptop so that it can be accessed easily anytime and anywhere (Suhartanto 2016). In addition, this website does not take up a lot of memory because it does not need to be downloaded first. Researchers developed this website to create practical media innovations for article writing skills training for teachers. The results of the development of a scientific article training website are presented in images that display features for users.

The training website for making scientific articles based on CAR in the learning process provides an opportunity for teachers to learn more about writing. This allows teachers to always continue to work for the sake of better education .(Jumrawarsi & Suhaili 2020). Article writing training can significantly affect teacher motivation and quality (Nurgiansah et al. 2021). Research is the higher order thinking skills needed to develop and support academic, professional, and student success. Teachers will be ready to face the era of global competition; Indonesia also needs to prepare reliable human resources starting with the quality of its teachers (Subekt et al. 2017).

The continuous professional development above and the implementation of reflective actions in one of the professional competencies of teachers must be improved. The motivation that drives teachers in the environment to take part in training activities in the form of a website is a matter of rank even though there are participants who take part in training not with rank motivation, namely teachers from private schools (Arta 2019). Common motivation is because a teacher's career path will determine income and an increase in income becomes a teacher's motivation to work professionally (Junaid et al. 2020). Salary and allowance factors are one of the factors that determine teacher motivation in sustainable professional development to improve class quality (Sagala et al. 2019).

If it refers to the teacher's obstacles in writing scientific papers, as stated, solving the problem is from the teacher himself. Weak writing culture is also linear with weak reading culture (Sumartini et al. 2019). Therefore, to write a teacher must also have a high reading culture. The teacher's low reading interest is one of the main problems in the aspect of teacher pedagogical competence (Susilowati 2018). Training activities support writing skills, especially if notes are made after reading, so that writing activities are efficient and the notes over time can be as considered complete as documents (Gunawan et al. 2018).

In such an innovation, a teacher with other teachers can discuss to find and formulate problems and obstacles in the classroom (Susilowati 2018). Thus, teachers and their friends can conduct CAR in a collaborative and solution-oriented manner. From here there will be awareness of the possibility of many problems that are made during the teaching and learning process. If a teacher is willing to do CAR and make scientific articles using websites with other teachers, there are many benefits in increasing knowledge (Nurhayati et al. 2018). Scientific writing is increasingly needed by teachers in the future. Collaborative CAR will be able to offer broad opportunities for the creation of written works while teaching more qualified experts (Daud & Kaleka, 2019).

3.2 *Website development features design, writing, and publishing scientific articles*

The Smart Teacher website, as one of the digital websites in the Website version, has several features that support learning and training of scientific articles including designing, writing, publishing, consulting and more. This website can be used to support the learning of scientific articles and CAR independently by the teacher. Especially during this pandemic, limiting space for movement and not being able to gather. This platform accompanies the teacher's activities before, after, and even during the implementation of CAR and helps in making scientific articles from research results. Before carrying out CAR, you can read and understand the steps that need to be completed. In addition, teachers can see learning videos that have been made along with examples of drafts and check lists before carrying out CAR activities (identify problems, formulate problems, plan actions, implement actions, collect data, manage data).

The design feature is the right feature to help teachers before making CAR. This feature provides practical guidance and practice. Teachers are directed to understand various CAR designs. The following are the steps that can be taken by the teacher and the desired material is selected: (1) problem identification, namely the initial observation stage of a case or learning situation in the classroom; (2) problem formulation, namely formulating what you want to improve or what might be developed as a new skill by means of a new solution; (3) formulation of the action plan in action research is not a difference or relationship hypothesis, but an action hypothesis. The action hypothesis formulation contains the proposed action to produce the desired improvement. After that, a clear description of the research procedure is to be carried out. The object, time, and duration of the action, as well as the research location are clearly stated. Procedures should be detailed from planning, implementation of action, observation, evaluation-reflection, which are recycled or cyclical. (4) Determination of steps for data collection in the form of procedures or techniques that include what aspects will be covered to obtain data, such as observations, interviews, questionnaires, and at the same time compiling the necessary instruments. (5) At this stage of data collection, researchers do things such as learning, observation, interviews, or other activities in accordance with the steps of data collection that have been determined previously. There are many ways to obtain data. (6) Data Processing and Analysis: At this stage, the researcher begins to process the data according to the steps that have been set in the determination of the data processing steps. The data are sorted, tabulated, graphed, or entered into formulas, then the processed data are analyzed, interpreted, and finally concluded.

Based on Susilowati's research (2018), CAR is a strategic way for educators to improve and or improve educational services for educators in the context of classroom learning. Practical steps for implementing CAR can be described. Therefore, this paper focuses on the main activities, such as: (1) planning, (2) acting, (3) observing, and (4) reflecting (Susilowati, 2018). The final report on CAR can be used to improve the learning process in the classroom, recommend learning models, and furthermore increase the researcher's credit score.

The writing feature is a follow-up feature of the planning activities that have been carried out. The writing feature provides information to write the results of CAR into a scientific article. The writing feature will display learning materials and videos to the teacher. The writing steps in the writing feature are: (1) making the title, the first part of the scientific article that is read and is the identity that represents the content of a scientific article. A title can make a first impression on people who see it. So it is necessary to have suggestions and input from other teachers; (2) abstract, short writing and providing a complete explanation of the contents of scientific articles. In this section, good, informative, and descriptive abstract tips will be given; (3) the method of procedure or rules used to carry out research. Methods that are structured systematically and scientifically, so that they become rules that must be carried out in conducting research; (4) results, the results section only describes the research results or raw data; (5) discussion, the data is described about the discussion of the analysis of research results displayed in the form of graphs, images, or tables, and equipped with data analysis in the form of text descriptions so that they are easy to understand; (6) conclusions, answers to the questions contained in the formulation of the problem or research question. The number of conclusions must be adjusted to the number of problem formulations that have been described in the introduction to scientific articles; (7) bibliography, a section containing reference sources or reference sources used by the author to cite literature as material for scientific articles. In this section, the teacher will get information related to the use of automatic citations, namely Mendeley or Zotero.

The publication feature is a feature that helps teachers find SINTA's indexed journals according to research topics that are made into scientific articles. The Smart Teacher website will link directly to the official website of the existing journal. This website also has data for grouping journals into selected topics. If the teacher clicks on the journal website according to the chosen topic, then the teacher can see the technicalities for publication and download the abstract and full paper templates that you want to write. Therefore, teachers will get easier access to publish scientific articles that they have made.

The consultation feature helps teachers consult regarding CAR that is or will be carried out. Teachers can send messages to the expert team if they encounter confusion or want to discuss. In addition, other teachers can provide input related to the work being made so as to create a good collaborative ecosystem between fellow teachers. Teachers can also consult related scientific articles to be published in journals and consult comments from the intended journal reviewers,

According to Sumartini's research (2019), mentoring and consultation in writing scientific papers so that teachers are able to make scientific works properly and correctly (Sumartini et al. 2019). Open access to interactive materials and information through the network opens up opportunities for collaboration between teachers, other teachers, principals, lecturers, and a team of experts (Fitriyadi 2013). The about feature provides information related to the website, the function of each feature on the website, so that teachers do not feel confused in using the website that has been created. This website can be used as a system that makes it easier for teachers to carry out CAR and increase teacher productivity in making scientific articles.

3.3 *Validation result*

At the validation stage, two lecturers at the State University of Malang, namely learning media experts were subjects. The validation stage ensures that the learning media developed meets the feasibility of being used by students based on the opinions of experts. In addition, it is in accordance with the stages in the preparation of scientific articles based on CAR. At this validation stage, what is assessed is in terms of the media. The instrument used is a questionnaire material validation sheet and a media validation questionnaire sheet which contains a statement of aspects adapted to a good media reference (Rokhim et al. 2020).

Based on the results of the media validator, a value of 78.3% is obtained, which means that the developed media is classified as valid because the assessment points of the validation results have reached a number in the average value range of 61–80 based on the validation criteria table (Arikunto 2010) and the products produced in the research, the design of the media display has met the requirements in terms of media format, language, media content, graphics, and aspects of media presence, so that the product can be used. This Smart Teacher website product will continue to be revised in accordance with the suggestions and comments that have been given by the validator in the media validation questionnaire to improve the media components developed. Comments and suggestions include the addition of moving images or videos in this web user guide because the average web user comes from the general public. Guidebooks on the web can be made in the form of videos and uploaded on YouTube. In addition, the revision of this media product was also carried out to achieve the very valid category of learning media for teachers. These comments and suggestions help researchers in perfecting the developed learning media products.

3.4 *Website developed for online learning media for teachers*

The website platform can be operated using an Android smartphone or laptop connected to a browser. The advantage of a website that can facilitate online learning because it can be used as a synchronous and asynchronous communication tool anytime and anywhere using a cellphone or PC. This advantage can increase teachers' understanding of CAR and making scientific articles as well as being flexible in their learning. The website platform allows teachers to communicate and collaborate with other teachers or a team of experts to create scientific articles based on CAR. The urgency of online learning includes 1) the Covid-19 pandemic, which has led all activities, including continuous professional development training activities for teachers, to be carried out without face to face interaction, 2) easy access to materials and collaborative arenas, and 3) demands for the use of technology in education that must be controlled by the teacher. With this website platform, teachers can carry out training activities efficiently.

4 CONCLUSION

The creation of a Smart Teacher website has shown that website design, writing, and publication of CAR-based scientific articles has five main menus, namely planning, writing, publishing, consulting, and about. The design feature is the right feature to help teachers before making CAR. The writing feature is a follow-up feature of the planning activities that have been carried out to write the results of CAR into a scientific article. The publication feature is a feature that helps teachers to find SINTA's indexed journals according to research topics that are made into scientific articles. The consultation feature is a feature that helps teachers to consult regarding CAR that will or is being carried out. The about feature is a feature that provides information related to the website, the function of each feature on the website. This website can improve the pedagogic competence of teachers. The media validator obtained a value of 78.3%, which means that the media developed is classified as valid. This score indicates that the virtual laboratory is very valid on material validation and valid on media validation. This shows that in terms of readability, the media can be understood clearly and has used the right choice of words. Weaknesses on the website can be minimized by policy, commitment, and further product development so that it can be focused on a forum for sustainable teacher professional development. This is supported by gains.

REFERENCES

Aisyah, E. N., & Mahanani, P. (2017). Pelatihan Menulisan Artikel Ilmiah Bagi Guru Sekolah Dasar Dan Taman Kanak-Kanak Kecamatan Tajinan Kabupaten Malang. *ABDIMAS PEDAGOGI, 1*(1), 22–26.

Alan, Ü., & Atalay Kabasakal, K. (2020). Effect of number of response options on the psychometric properties of Likert-type scales used with children. *Studies in Educational Evaluation, 66*(October 2019), 100895.

Arta, K. S. (2019). Pelatihan Penulisan Artikel Untuk Publikasi Di Jurnal Ilmiah Untuk Meningkatkan Profesionalisme Bagi Guru-Guru Di Kecamatan Kubutambahan Kabupaten Buleleng. *Acarya Pustaka, 5*(2), 17.

Basturk, S. B., Dancer, C. E. J., & McNally, T. (2020). Journal Pre. *Pharmacological Research*, 104743.

Christoforidou, M., & Kyriakides, L. (2021). Developing teacher assessment skills: The impact of the dynamic approach to teacher professional development. *Studies in Educational Evaluation, 70*(November 2020), 101051.

Darmadi, H. (2015). Tugas, Peran, Kompetensi, Dan Tanggung Jawab Menjadi Guru Profesional. *Jurnal Edukasi, 13*(2), 161–174.

Daud, M. H., & Kaleka, M. (2019). Analisis Kesulitan Bagi Guru Ipa Untuk Melakukan Penelitian Tindakan Kelas (PTK). *OPTIKA: Jurnal Pendidikan Fisika, 3*(2), 36–42.

Fatkuroji. (2019). *Revitalisasi Sumberdaya Manusia Melalui Pengembangan Kebudayaan Kabupaten Grobogan. 12*(2), 13–26.

Fitriyadi, H. (2013). Integrasi Teknologi Informasi Komunikasi Dalam Pendidikan: Potensi Manfaat, Masyarakat Berbasis Pengetahuan, Pendidikan Nilai, Strategi Implementasi Dan Pengembangan Profesional. *Jurnal Pendidikan Teknologi Dan Kejuruan, 21*(3), 259–615.

Gunawan, I., Triwiyanto, T., & Kusumaningrum, D. E. (2018). Pendampingan Penulisan Artikel Ilmiah bagi Para Guru Sekolah Menengah Pertama. *Abdimas Pedagogi: Jurnal Ilmiah Pengabdian Kepada Masyarakat, 1*(2), 128–135.

Jumrawarsi, & Suhaili, N. (2020). Peran Seorang Guru Dalam Menciptakan Lingkungan Belajar Yang Kondusif. *Ensiklopedia Education Review, 2*(3), 50–54.

Junaid, R., Baharuddin, M. R., & Ramadhana, M. A. (2020). Bimbingan Teknis Penulisan Artikel Ilmiah Bagi Guru-Guru Smp Negeri 8 Palopo, Sulawesi Selatan. *Abdimas Siliwangi, 03*(01), 49–59.

Kuncoro, R. B. (2014). Pembuatan Website Tempat Parawisata Rumah Dome New Nglepen. *Speed Journal – Sentra Penelitian Engineering Dan Edukasi, 11*(4), 33–40.

Kurnia, T. D., Lati, C., Fauziah, H., & Trihanton, A. (2019). Model ADDIE Untuk Pengembangan Bahan Ajar Berbasis Kemampuan Pemecahan Masalah Berbantuan 3D. *Seminar Nasional Pendidikan Matematika, 1*(1), 516–525.

Muchson, M., Munzil, Setiawan, N. C. E., Sari, M. E. F. S., Novitasari, S., & Rokhim, D. A. (2021). Program Pembinaan Pengembangan Media Pembelajaran Bagi Guru Kimia MA / SMA Pada MGMP Kimia Kabupaten Mojokerto Berbasis IoT Learning Media Development Training Program for Chemistry Teachers of MA / SMA in the Chemistry MGMP of Mojokerto Regency Based on. *Jurnal Pengabdian Kepada Masyarakat, 5*(3), 420–431.

Nurgiansah, T. H., Pratama, F. F., & Iman Nurchotimah, A. S. (2021). Penelitian Tindakan Kelas Dalam Pendidikan Kewarganegaraan. *Jurnal Pendidikan PKN (Pancasila Dan Kewarganegaraan), 2*(1), 10.

Nurhayati, T., Nurunnisa, E. C., & Husni. (2018). Upaya Meningkatkan Kemampuan Membaca Al-Qur'an Anak Usia Dini Melalui Penerapan Metode Iqra' (Penelitian Tindakan Kelas di Raudhatul Athfal Daarul Hikmah Kecamatan Cijeungjing Kabupaten Ciamis). *Tarbiyat Al-Aulad, 3*(1), 1–6.

Riyadi, S. (2016). Kebijakan Pengembangan Profesionalisme Guru PNS. *JKMP, 4*(2), 117–234.

Rokhim, D. A., Wulandari, I. A. I., & Alifah, L. N. (2020). Analisis dampak anjuran pemerintah terhadap belajar di rumah bagi pelaku pendidikan. *JAMP: Jurnal Adminitrasi Dan Manajemen Pendidikan, 3*(3), 216–233.

Sagala, D., Supriyono, S., Sutardi, D., & Prihanani, P. (2019). Pelatihan dan Pendampingan Penulisan Artikel Ilmiah pada Guru Sekolah Dasar di Kecamatan Air Periukan Kabupaten Seluma Provinsi Bengkulu. *Abdihaz: Jurnal Ilmiah Pengabdian Pada Masyarakat, 1*(1), 1.

Schuster, J., Hartmann, U., & Kolleck, N. (2021). Teacher collaboration networks as a function of type of collaboration and schools' structural environment. *Teaching and Teacher Education, 103*, 103372.

Subekt, H., Taufiq, M., Susilo, H., Ibrohim, I., & Suwono, H. (2017). Mengembangkan Literasi Informasi Melalui Belajar Berbasis Kehidupan Terintegrasi Stem Untuk Menyiapkan Calon Guru Sains Dalam Menghadapi Era Revolusi Industri 4.0: Revieu Literatur. *Education and Human Development Journal, 3*(1), 81–90.

Suhartanto, M. (2016). Pembuatan Website Sekolah Menengah Pertama Negeri 3 Delanggu Dengan Menggunakan Php Dan Mysql. *Sentra Penelitian Engineering Dan Edukasi, 4*(1), 1–8.

Sumartini, S., Mulyani, M., & Nugroho, B. A. (2019). Workshop Penulisan Karya Ilmiah Bagi Guru Sekolah Dasar di Kabupaten Demak. *Jurnal Puruhita, 1*(1), 54–59.

Susilowati, D. (2018). Penelitian Tindakan Kelas (Ptk) Solusi Alternatif Problematika Pembelajaran. *Jurnal Ilmiah Edunomika, 2*(01), 36–46. https://doi.org/10.29040/jie.v2i01.175

Improving Assessment and Evaluation Strategies on Online Learning – Wonorahardjo, Karmina & Habiddin (eds)

Implementation of the discovery learning model in regional economics

F. Rahmawati*, I. Mukhlis & E. Yusida
Universitas Negeri Malang, Indonesia

ABSTRACT: In regional economics courses, students' interest in learning is low, indicated by the completion of tasks given by lecturers to students, which takes a long time and the results are not optimal. In addition, student learning outcomes related to regional economic subject matter are also low. The method used in this research is a descriptive qualitative approach with classroom action research, which is carried out in two cycles, each cycle is carried out three times. The implementation of this model is carried out in four stages, namely planning, action, observation, and reflection. The results showed that the application of the discovery learning model in two cycles could increase students' interest in learning effectively and could also improve student learning outcomes. This study recommends the need to combine discovery learning models with project-based learning in finding leading economic sectors in regional economics courses.

Keywords: discovery learning model, regional economics, students interest, student learning outcomes

1 INTRODUCTION

Regional economics is a branch of economics that in its discussion includes elements of potential differences from one region to another. Regional economics does not discuss individual activities, but discusses and analyzes a region, either in whole or in part with its diverse potential, as well as how to regulate the economic growth policy of a region. This regional economy is very important to study, because the economic potential of each region is different, so the policies applied are also different depending on the characteristics of each region. This is in line with the statement of Levine (1997) and Wennekers et al. (2005), which says that different levels of regional economic development create a difference in regulating or making policies in a region.

Knowledge of regional economics is one of the most important things for students of the Department of Development Economics, because regional economics is one of the branches of economics, so that its knowledge is a science that must be studied by students of the Department of Development Economics. Regional economics is a branch of economics which in its discussion includes elements of the differences in the potential of a region to other regions (Tarigan 2005). According to Florida et al. (2008) the drivers of economic growth and development are skilled and educated people, so economics students are obliged to study regional economics to realize better economic growth and development. Therefore, for students to be richer in knowledge both in terms of cognitive or experience, it is necessary to use a discovery-based learning model or commonly called the Discovery Learning Model in regional economics courses, with the aim that students become more active, can think critically, and gain rich experience. Several studies and theories that support the use of Discovery Learning Model can increase students' participation, activeness, and thinking skills, including research conducted by Wulandari et al. (2015), Rosarina (2016), and Arends (2012).

The need for the use of Discovery Learning Model in regional economics course is that previously students' interest in learning was low, and learning outcomes were still low, which could be seen

*Corresponding Author

 DOI 10.1201/9781003261346-7

from the results of the pre-test before using the Discovery Learning Model. Therefore, the use of Discovery Learning Model in regional economics learning activities is expected to increase student interest in regional economics courses, and can improve students' learning outcomes. In addition, what is expected from the use of this model in regional economics courses is that the quality of graduates from the Department of Development Economics is increased, which will be evident through making decisions or making policies.

2 METHODS

This research is a research using a qualitative approach with the type of classroom action research. This classroom action research aims to improve the learning process in the classroom so that it can achieve the expected learning objectives, to increase student interest in regional economics courses and to improve student learning outcomes. This research was conducted in two cycles, each cycle consisting of three meetings. The research consists of four stages in each cycle. The stages of classroom action research are planning, action, observation, and reflection. Cycle I consists of these four stages; the same thing is carried out in cycle II, with the aim of addressing deficiencies that have been carried out in cycle I. The subjects in this study were students who were participants in the regional economics course in the Even Semester of the 2020/2021 academic year in the Undergraduate Development Economics Study Program, Faculty of Economics, Universitas Negeri Malang. The instruments used are observation sheets regarding the success of the lecturer's actions, test questions, affective assessment observation sheets, and field notes. Data collection techniques use test and non-test techniques. Test technique is used to get learning outcomes in the cognitive domain. The non-test technique is in the form of observations made to obtain data on the percentage of the success rate of the researcher's actions in applying the Discovery Learning Model.

3 RESULTS AND DISCUSSION

3.1 *Application of the discovery learning model in regional economics course*

Before applying the Discovery Learning Model to the regional economics course, it was found that: 1) students did not understand the material that had been delivered by the lecturer, 2) students did not pay attention to the material explained by the teacher; this can be proven when there are questions. Students are less able to answer questions correctly. appropriate, and lack of enthusiasm from students, 3) students are less active during discussions, 4) students are less active in asking questions, 5) students are also less than optimal in doing the assignments given by the lecturer even though they have been given a long time, 6) learning outcomes in students are low; this is obtained from the results of the pre-test before using the Discovery Learning Model. This happens because the previous regional economics course used a learning model that tends to be centered on educators so as to make students passive in learning activities, this is what makes students less interested in regional economics courses. Students tend to be passive because of the lack of student understanding of the concept of the material that has been given by the lecturer, the learning model that is not suitable so that it has an impact on the ability to think critically, interest and student learning outcomes are lacking. This is in line with Becker and Watts (2001), who state that students' interest will increase if they use a variety of learning models, and invite students to be involved in speaking, writing, discovering, and others. Therefore, in completing the tasks given by the lecturer, students cannot work optimally as expected by the lecturer. Even in the tests conducted before using the Discovery Learning Model, student learning outcomes are low. In addition, before using the Discovery Learning Model in regional economics courses, students are also afraid or hesitant to express their ideas and ideas, and even have to be appointed first before expressing their opinions. Because there are several problems as described above, teaching lecturers develop new strategies so that students are much more active in learning activities for regional economics courses, and also so that student learning outcomes also progress or improve, therefore teaching lecturers use the Discovery Learning Model in the regional economics course.

After using the Discovery Learning Model in the regional economics course the results were found to be more effective, students became more active than before, students answered more frequently asked questions, and often responded or gave opinions when there was a discussion forum. In addition, students become more enthusiastic during learning activities and can answer teacher's questions correctly even when there are assignments. Students work with maximum results than before. Students complete assignments on time and with better and maximum work results than before. This is in line with Tran (2014) who said that the Discovery Learning Model is more effective than the traditional learning model. In addition, by using this learning model in regional economics courses, students can find regional superior sectors according to the themes that have been determined by the teaching lecturers. Student knowledge increases in line with the findings. This is in line with research conducted by Mukherjee (2015), who states that students increase their knowledge more, because they find this knowledge. According to Mukherjee (2015) the expectations from discovery learning is that students have a better understanding of the knowledge they have discovered themselves.

3.2 *Increasing student interest and learning outcomes in regional economics courses*

3.2.1 Interest to learn

Students' learning interest can be identified through a questionnaire distributed to students using a Google Form given to 25 undergraduate students of Development Economics at Universitas Negeri Malang, who are taking regional economics courses in the even semester of the 2020/2021 academic year. The questionnaire contains a number of questions and statements that are used to measure student interest in the regional economics course. The questionnaire distributed to several respondents also contained guidelines for filling out, so that the results of the questionnaire that were filled out could describe the actual situation. The results of the questionnaire distributed to several respondents, will then be analyzed and produce research results related to student interest in regional economics courses after using the Discovery Learning Model. The following are the results of research on student interest in learning before and after using the Discovery Learning Model:

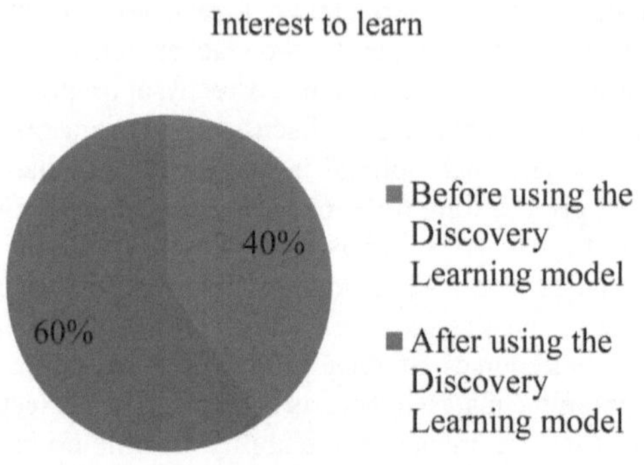

Figure 1. Student interest in regional economics.

Based on the diagram in Figure 1 above, it shows that student interest in regional economics courses before using the Discovery Learning Model is 40%, and interest after using the Discovery Learning Model increases to 60%. It can be concluded that student interest in regional economics courses after using the Discovery Learning Model can increase interest in regional economics courses by 20%. In line with Shieh & Yu's (2016) statement, which states that learning using the Discovery Learning Model makes students active in participating in learning activities, and students can build their own knowledge, this can increase students' understanding of the relevant learning material. Other research that supports that the use of the Discovery Learning Model can increase students' interest in learning is a research conducted by Puspitadewi et al. (2016) and Rustam et al. (2019).

3.2.2 Learning outcomes

Learning research uses the Discovery Learning Model in the regional economics course, in addition to researching student interest in learning, also to determine student learning outcomes. This study of learning outcomes uses written test instruments and assignments given by the teacher to students. Below are the student learning outcomes in the regional economics course before using the Discovery Learning Model and after using the Discover Learning Model.

Table 1. Student learning outcomes in pre-cycle regional economics course, Cycle I and Cycle II.

Indicator	Pre-Cycle	Cycle I	Cycle II
Total of students	25	25	25
Total value	1,550	1,845	2,075
Average	62	73,8	83
The highest score	75	85	95
Lowest value	40	55	70
Percentage of students complete	32%	64%	92%
Percentage of students who did not complete	68%	36%	8%

Students' interest in studying regional economics will also improve student learning outcomes. This is in accordance with the results of the study, which found that when regional economic learning used the Discovery Learning Model, student interest in learning increased and student learning outcomes also increased. Interest in learning can be seen from the observations of observers and field notes, that when learning regional economics uses the discovery learning model, students become more active than before using the discovery learning model. In addition, learning outcomes also showed a significant increase in cycle I and cycle II. Cycle I and cycle II were both carried out for three meetings. The percentage of students who completed before using the Discovery Learning Model was 32%, after using the Discovery Learning Model increased to 64% in the first cycle and increased to 92% in the second cycle. From the presentation of the data above, it can be concluded that student learning outcomes increased after using the Discovery Learning Model. This is in line with the results of research conducted by Yuliani & Saragih (2015), which states that learning outcomes increase critical thinking skills when learning using the Discovery Learning Model. This statement is also supported by the research by In'am & Hajar (2017), which says that the application of the Discovery Learning Model of learning outcomes for students is improving. Other studies that support the use of Discovery Learning Model for improving student learning outcomes are Yuliana et al. (2017) and Batubara (2014).

4 CONCLUSION

Based on the discussion, it can be concluded that the use of Discovery Learning Model in two cycles in regional economics courses has proven to be very effective, because it makes students more active and creative, and students' interest in learning about the regional economy is increasing. This can be seen from the enthusiasm of students in asking and answering questions, the enthusiasm of students in completing the tasks given by the lecturer, and creativity and critical thinking skills can be seen from the outputs of the Discovery Learning Model in the form of modules and learning videos. From the explanation above, it shows that the learning outcomes and the tasks carried out by students are much better after using the Discovery Learning Model, than before using the model whose work results are less than optimal. Learning outcomes also showed a significant increase in each cycle when using the Discovery Learning Model. Increasing student interest in learning in the regional economics course also improves student learning outcomes, in terms of cognitive, affective, and psychomotor aspects. Based on the research results, the researcher also recommends the need to combine the Discovery Learning Model with the Project-based Learning Model in finding the leading economic sector in regional economics courses.

REFERENCES

Arends, R I. 2012. *Learning to Teach ninth edition.* New York : McGraw-Hill.

Batubara, Anni Erlina. 2014. *The Influence of Inquiry and Discovery Learning Strategies on Critical Thinking Ability and Biology Learning Outcomes Of Students on The Topic of Biotechnology at MAN 1 Padangsidimpuan.* Thesis of the Medan State University Postgraduate. (Unpublished).

Becker, W. E., & Watts, M. 2001. *Teaching Methods in U.S. Undergraduate Economics Courses. The Journal of Economic Education, 32(3), 269–279.* doi:10.1080/00220480109596108

Florida, R., Mellander, C., & Stolarick, K. 2008. *Inside the black box of regional development–human capital, the creative class and tolerance. Journal of Economic Geography, 8(5), 615–649.* doi:10.1093/jeg/lbn023

In'am, A., & Hajar, S. 2017. *Learning Geometry through Discovery Learning Using a Scientific Approach.* International Journal of Instruction,10(1), 55–70. https://doi.org/10.12973/iji.2017.1014a

Levine, R., .1997. *Financial development and economic growth: views and agenda.* Journal of Economic Literature XXXV, 688–726.

Mukherjee, A. 2015. *Effective Use of Discovery Learning to Improve Understanding of Factors That Affect Quality.* Journal of Education for Business, 90(8), 413–419.

Puspitadewi, R., Saputro, ANC., & Ashadi.2016. *Application of the Discovery Learning Model to Increase Students Interest and Learning Achievement in Solubility Materials and Solubility Product Class XI MIA 3 Even Semesters of SMAN 1 Teras 2015/2016 Academic Year.* Chemistry Education Journal. Vol 5 No.4 Tahun 2016.

Rosarina, G,. Sudin, A., Sujana, A. 2016 . *Application of the Discovery Learning Model to Improve Student Learning Outcomes on Material Changes in Objects.* Education Journal PGSD UPI.

Rustam, S., Saragih, S., & Hastarudin. 2019. *Improving Students Mathematical Problem Solving Ability and Self Efficacy Through Guided Discovery Learning in Local Cilture Context.* International Electronic Journal Of Mathematics Education.

Shieh, C. J., & Yu, L. A. 2016. *Study on Information Technology Integrated Guided Discovery Instruction toward Students Learning Achievement and Learning Retention.* Eurasia Journal of Mathematics, Science & Technology Education, 12(4), 833–842. https://doi.org/10.12973/eurasia.2015.1554a

Tarigan, Robinson. 2005. *Regional Economics Theory and Application (Revised Edition).* Jakarta: PT Bumi Aksara

Tran, Trung. 2014. *Discovery Learning with the Help of the Geogebra Dynamic Geometry Software. International Journal of Learning, Teaching and Educational Research (IJLTER).* Vol. 7 (1), pp. 44–5.

Wennekers, S., Wennekers, A., Thurik, R., Reynolds, P., .2005. *Nascent entrepreneurship and the level of economic development.* Small Business Economics 24, 293–309.

Wulandari,Yun Ismi, Sunarto dan Salman Alfarisy Totalia. 2015. *Implementation of the Discovery Learning Model With a Scientific approach to Improve Critical Thinking ability and Student Learning Outcomes of Economics Subject Class XI IIs SMA Negeri 6 Sukrakarta Academic Year 2014/2015.* FKIP Universitas Sebelas Maret Surakarta, page 8.

Yuliana, Tasari & Wijayanti, S. 2017. *The Effectiveness Of Guided Discovery Learning to teach Integral Calculus For the Mathematics Students of Mathematics Education Widya Dharma University.* Journal Of Mathematics Education. Volume 6 No 1 2017.

Yuliani, K., & Saragih, S. 2015. *The Development of Learning Devices Based Guided Discovery Model to Improve Understanding Concept and Critical Thinking Mathematically Ability of Students at Islamic Junior High School of Medan.* Journal of Education and Practice, 6(24), 116–128.

Improving Assessment and Evaluation Strategies on Online Learning – Wonorahardjo, Karmina & Habiddin (eds)

Religiosity and nomophobia among undergraduate student: The moderating role of self-control

N.N. Afifah & H.E. Wijaya*
Universitas Islam Indonesia, Indonesia

ABSTRACT: This research aims to investigate the moderating role of self-control in the relationship between religiosity and nomophobia among students. A convenient sample of 288 undergraduate students enrolled in this study in 2019. We administered three self-report instruments to collect data, the No Mobile Phone Phobia Questionnaire, religiosity scale, and brief self-control scale. First, we applied Pearson correlation analysis to find out the relationship between the variables. The results showed there was a negative correlation between religiosity and nomophobia, as well as a negative correlation between self-control and nomophobia. Analysis results also point out that there is a positive relationship between religiosity and self-control. Furthermore, moderator analysis indicated that self-control could moderate the relationship between religiosity and nomophobia. These findings underlined the role of self-control as a protective factor when students were prone to nomophobia.

Keywords: Self Control, Religiosity, Nomophobia, Undergraduate student

1 INTRODUCTION

Smartphones or gadgets have become a part of our daily life. Global data shows that by the end of 2020, near half of the world's population would have a smartphone, which is 46.45% (O'Dea, 2021). The trend of smartphone users is growing each year. Young children to adults have the same urge to use smartphones. Moreover, the various features offered by smartphones like social media, online shopping, online learning, or entertainment, could make users addicted. It is not surprising currently, Indonesia has to reach smartphone users no less than 89% of the total population, which is about 169 million people (Hanum 2021).

Nowadays, smartphones not only make communication easy but also facilitate our economic, academic, and social life activities. However, the use of smartphones that exceed reasonable limits reportedly has a lot of negative impact on their users. The results of a meta-analysis study by Sunday, Adesope, and Maarhuis (2021) showed a negative impact of excessive smartphone use on academic performance. In addition, excessive use of smartphones also has a detrimental effect on mental health, such as obsession-compulsion, interpersonal sensitivity (Gonçalves et al. 2020), healthy lifestyle (Alosaimi et al; 2016), physical and psychological health (Davey & Davey 2014), as well as spiritual health (Buctot et al. 2020). In the work context, dependence on smartphones could also lead to distress (Tams et al. 2018).

The excessive use of smartphones or gadgets could make an individual dependent on his smartphone. The phenomenon of not being able to separate from a smartphone in various situations, whether working, studying, or talking to others, is known as nomophobia. The word nomophobia stands for no mobile phone phobia, which is related to behavior in using gadgets. Nomophobia itself is categorized as a situational phobia, where a person is frightened, anxious, and experiences physical or emotional sensations when far away or cannot access his smartphone (King & Silva 2014). Yildirim and Correia (2015) explain that nomophobia is a new phobia related to access

*Corresponding Author

DOI 10.1201/9781003261346-8

to lost information, lost interactions, and lost communication. This is why nomophobia leads to anxiety (Tams et al. 2018).

Bragazzi and Puente (2014) have proposed six indicators of nomophobia. First, the person is constantly feeling urged to use smartphone for a long time. Second, feeling anxious or worried when the smartphone is not nearby if lost, left behind, the battery runs out, or there is no signal. Third, keep checking the phone's screen as worrying if there are any incoming messages. Fourth, keep turning on the smartphone continuously throughout the day until sleep. Fifth, having a little face-to-face interaction with others because the person prefers to contact via mobile phone. Sixth, having high costs as a result of excessive mobile phone use.

The prevalence of nomophobia is different in various places. Adolescents in the UK reported having problems using a smartphone as much as 10% of the total sample of 1,026 adolescents (Lopez-Fernandez, Honrubia-Serrano, Freixa-Blanxart, &Gibson, 2014). The excessive use of smartphones was also found in Indian adolescents. A meta-analysis of 45 smartphone addiction studies conducted by Davey and Davey (2014) showed 39%–44% of Indian adolescents were addicted to smartphones. In line with that study, Alosaimi et al. (2016) pointed out that 27.2% of the 2,367 students in Saudi spend more than eight hours using a smartphone each day.

Similar findings were also found in Cha and Seo's study (2018) in South Korea. They observed that 30.9% of high school students were at risk of having smartphone addiction. According to Bragazzi, Re, and Zerbetto's research (2019), students in Italy experienced nomophobia. Of them, 51.1% reported having low levels, 41.4% moderate, and 7.4% having severe levels. There is a tendency that students with high level of nomophobia would use inappropriate coping strategies when facing pressures, such as blaming themselves, denying, seeking distractions, or staying away from problems.

Harish and Bharath's (2018) research in India showed that of the 418 students the percentage of usage in students was moderate level (50.4%), followed by low level (36.1%), and high level (13.5%). Similar to those findings, Buctot, Kim, and Kim's (2020) study of 1,447 adolescents in the Philippines also found having nomophobia. Around 12.4% reported had low nomophobia levels, 63.2% moderate levels, and 23.8% high levels.

Alongside those findings, research conducted at Sultan Qaboos University in Oman has shown few cases of nomophobia. In more detail, 20% had low-level nomophobia, 65% moderate, and 15% severe (Qutishat et al. 2020). Meanwhile, exploratory studies in Pekanbaru, Indonesia, showed that 12% of 260 samples indicated having had a high level of nomophobia and 78% had a moderate level (Rezki et al. 2018).

One of the factors related to nomophobia is the ability of individuals to take control of their actions (Asih & Fauziah 2018; Busch & McCarthy;021). Self-control is one's ability to direct behavior based on particular standards, such as rules, values, and morals in society, which will lead to desirable behavior (Tangney et al. 2004). de Ridder et al.'s (2018) meta-analysis pointed out that people with high self-control not only are capable to inhibit unexpected behavior but also could promote desired behavior. Accordingly, students with high self-control students could avoid nomophobia.

Another relevant construct that correlates with nomophobia is religiosity. Some studies had considered religiosity as a protective factor against nomophobia (Fereday & Chair 2017; Shim 2019). Yonker, Schnabelrauch, and DeHaan's (2012) meta-analysis also highlighted the positive impact of religiosity on risky behaviors, depression, psychological well-being, self-esteem, and personality. As an addition, when individuals face obstacles or difficulties, religiosity could serve as a source of resilience (Reutter & Bigatti 2014). Therefore, individuals with high religiosity would be able to prevent nomophobia.

On the other hand, religiosity can affect student self-control (Watterson & Giesler 2012). According to McCullough and Willoughby (2009), religiosity could strengthen self-control because it facilitates individuals' awareness of being watched by God or feeling monitored by the community. Carter, Mccullough, and Carver (2012) explained that self-monitoring works as self-awareness in mediating the relationship between religiosity and self-control. Religious individuals tend to have beliefs always supervised by God. For this reason, they have to be careful when behaving or doing a good deed.

Research conducted on religiosity and self-control is abundant, as mentioned earlier. However, there is still little study that examines the relationship between religiosity and nomophobia. Moreover, finding out the relationship between religiosity, self-control, and nomophobia is still very limited. Based on that background, we attempt to find out whether self-control could moderate the relationship between religiosity and nomophobia. There are four hypotheses propose in this study: 1) Religiosity has a negative relationship with nomophobia; 2) Self-control has a negative relationship with nomophobia; 3) Religiosity has a positive relationship with self-control, and 4) Self-control would moderate the relationship between religiosity and nomophobia.

2 METHODS

2.1 *Participants*

Participants in this study were undergraduate students aged 18–24 years old. We took a convenient sample by distributing the questionnaire on social media. Two hundred and eighty-eight students filled out online questionnaires via a Google Form. They were 141 females (49%) and 147 males (51%), dominantly from a few private universities in Yogyakarta (75%) and the rest is several public universities (25%). This research was conducted from the first month of 2019 until mid-year.

2.2 *Measure*

Three questionnaires were employed to obtain data. They are the No Mobilephone Phobia Questionnaire (NMP-Q) scale, religiosity scale, and The Brief Self-Control scale. The NMP-Q was developed by Yildirim and Correia (2015). For this current research, we use NMP-Q which was previously translated by Saputra and Dewi (2018). This questionnaire consists of four dimensions: inability to communicate, loss of connectedness, inability to access information, and giving up convenience. The questionnaire has 20 items wherein internal reliability reported on Cronbach's alpha is 0.927 and a corrected item-total correlation is between 0.283 and 0.807.

The self-control scale used in this study is the Brief Self-Control Scale. It was developed by Tangney, Baumeister, and Boone (2004), which consists of five dimensions of self-control: self-discipline, non-impulsive action, habitual to live healthy, self-regulation, and consistency. The scale has been translated and used by Rahmantori and Wijaya (2018) with a Cronbach's alpha value of 0.803 and a coefficient of corrected item-total correlation ranging from 0.282 and 0.633. The total number of items is 13.

Religiosity was measured using the religiosity scale, which was translated by Batong and Uyun (2014). It was developed based on the five dimensions of Glock and Stark religiosity: belief, rituality, experience, knowledge, and experience or consequences. This religiosity scale consists of 28 items and seven questions. Cronbach's alpha reliability coefficient on this scale was 0.798, and the coefficient of the corrected item-total correlation religiosity scale ranges from 0.202 to 0.544.

2.3 *Data analysis*

We were using IBM SPSS Statistics 22.0 for the Windows program to analyze the data. To determine the relationship between the three variables, we used bivariate correlation and model 1 Process v.3.0 by Andrew F. Hayes to examine the moderator analysis.

3 RESULT AND DISCUSSION

3.1 *Result*

Before conducting correlation analysis, we examined the normality and the linearity of all variables. Religiosity, self-control, and nomophobia meet standard distribution assumptions. It was checked by visual inspection on histograms and Q-Q plots. The linearity test showed that the relationship between the variables was linear.

Table 1. The correlation of religiosity, self-control, and nomophobia.

Variables	1	2	3	M	SD
Religiosity	–			125.86	7.63
Self-control	0.61**	–		42.66	6.48
Nomophobia	−0.13*	−0.13*	–	61.63	14.58

*Significant correlation with $p < 0.05$; **$p < 0.01$

The first analysis showed that religiosity had a negative and significant relationship with nomophobia ($r = -0.13$; $p < 0.05$). The first hypothesis is accepted. The second hypothesis is also accepted, which showed that self-control has a negative and significant relationship with nomophobia ($r = -0.13$; $p < 0.05$). Correlation analysis on the third hypothesis showed that religiosity has a positive and significant relationship with self-control ($r = 0.61$; $p < 0.01$). Therefore, the third hypothesis is accepted. According to Cohen (1992), the correlation between religiosity and self-control is high. The relationship between religiosity and nomophobia, as well as self-control on nomophobia, is weak.

Moderator's analysis indicates that self-control explained a significant increase in nomophobia (R2 = 0.02, F(1, 284) = 6.18, $p < 0.05$). Thus, self-control was a significant moderator of the relationship between religiosity and nomophobia. The unstandardized simple slope for self-control can be seen in Figure 1. When the student has low self-control, the relationship between the two variables is weak. However, with high self-control, the relationship between the two variables

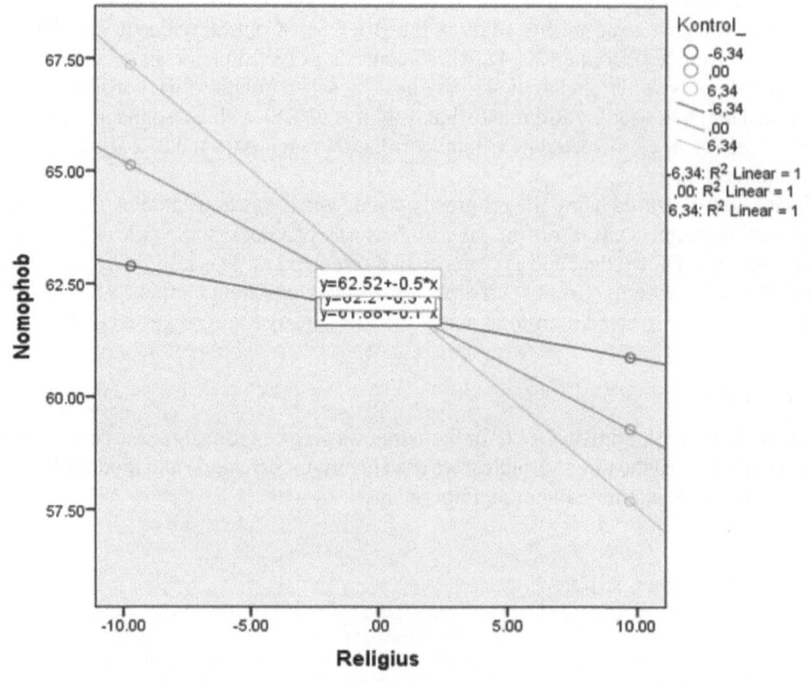

Figure 1. Graph of self-control analysis as moderator of the relationship between religiosity and nomophobia.

strengthens. In other words, high levels of self-control affect the relationship between religiosity and nomophobia.

3.2 *Discussion*

This study aims to find out the relationship between religiosity, self-control, and nomophobia among undergraduate students. In addition, we also want to test the self-control as a moderator of the relationship between religiosity and nomophobia. The analysis showed that the three variables have a significant correlation. Self-control can also act as a moderator of the relationship between the two variables.

The first results of this study corroborate Fereday and Chair's work (2017). They found religiosity was negatively correlated with nomophobia. Similarly, this study is also consistent with Shim (2019), which showed a negative correlation between smartphone addiction and spiritual wellbeing in Christian adolescents in South Korea. It also underlines Bragazzi, Re, and Zerbetto's (2019) findings, which showed indications of a relationship between religiosity and nomophobia. Although only in one aspect of nomophobia, loss of connection.

Geyer and Baumeister (2005) explained that religion could act as a source of virtue and value, with which an individual can control his attitude or behavior. According to him, religion has an essential function in providing standards or purposes of life, and there are provisions that are seen as good or bad. In addition, faith also motivates good act, in exchange for a reward or promise of a pleasant hereafter. Another function of religion is to facilitate individuals to monitor their behavior. Religion provides a foundation of belief in the supervision of God so that individuals avoid bad deeds or sins. Therefore, in this study, students with high degree of religiosity will direct themselves not to be fixated on smartphones by remembering the purpose or standards of kindness, such as not wasting time.

The second results showed a negative relationship between self-control and nomophobia. This output is in line with Asih and Fauzia's (2018) research, which showed that self-control could explain a 4.3% variance of variable nomophobia in Communication Science students at Diponegoro University. This study also corroborates Adiyaman, Mudjiran, and Afdal's studies (2020) that showed a negative correlation between self-control and smartphone addiction in students at Padang State University. In this context, self-control was able to explain 30% of smartphone addiction variance. It also underlines Servidio's research (2019), which yields a negative relationship between self-control and excessive smartphone use. Self-control is essentially an individual's capacity to change or override the tendency of self-dominant response or the ability to regulate behavior, emotions, or thoughts (de Ridder et al. 2018). Students who have high self-control can inhibit any temptation of using a smartphone and turn away into other better activities (such as doing tasks, studying, etc.). Therefore, students with high self-control can avoid nomophobia.

The third result showed that religiosity has a positive relationship with self-control. It confirmed Laird, Marks, and Marrero's study (2011), which stated that adolescent religiosity in America is negatively related to self-control. This argument also highlights Watterson and Giesler's experiments (2012), where religiosity and self-control are causal. Change in religiosity affects student self-control. Furthermore, according to Carter, Mccullough, and Carver (2012), the relationship between the two variables can be explained by self-monitoring. High religiosity students perceive that God or others are watching him; consequently, they become more aware or careful in acting. Therefore, self-control will also increase as the awareness of the supervision arise.

The last results showed that self-control could act as a moderator in the relationship between religiosity and nomophobia. The intensity of the relationship between the two variables is determined by the capability to control oneself. The results underlined Laird, Marks, and Marrero's findings (2011). They explained that religiosity could act as a promotive factor of self-control and play a protective factor in anti-social behavior. In this current study, high religiosity students would strengthen the ability to resist the temptation to use smartphones or override feelings of anxiety or worry when apart from their phones. Religiosity can serve as a factor to strengthen self-control and, at the same time, protect students from the risk of nomophobia.

Some shortcomings need to be of concern in this research. First, the participants in this study were limited. Mostly they came from private universities in Yogyakarta. Therefore, it is necessary to be careful to generalize on populations beyond that. Second, this study aims to find out the relationship between the three variables. Thus, it is a correlational study and cannot interpret causality. To evaluate whether religiosity is the cause of nomophobia, we need to conduct experimental research.

4 CONCLUSION

The findings of this study highlighted the role of religiosity and self-control toward nomophobia. It has a significant relationship, both with self-control and nomophobia in undergraduate students. Religiosity serves as a protective factor for students when interacting with gadgets or smartphones. That role will be more powerful when students have a high level of self-control. Self-control, in this case, may become a booster to enhance religiosity's effect on nomophobia. Therefore, it should be the responsibility of policymakers at the university level in this digital era to develop programs to strengthen religiosity and self-control capabilities. Student productivity can be optimized when it is not dependent on the presence of smartphones.

REFERENCES

Adiyatma, R., Mudjiran, M., & Afdal, A. (2020). The contribution of self-control towards student smartphone addiction. *SCHOULID: Indonesian Journal of School Counseling*, *5*(1), 45.

Alosaimi, F. D; Alyahya, H; Alshahwan, H; Al Mahyijari, N; & Shaik, S. A. (2016). Smartphone addiction among university students in Riyadh, Saudi Arabia. *Saudi Med J*; 37(6): 675–683.

Asih & Fauziah. (2018). Hubungan antara kontrol diri dengan kecemasan jauh dari smartphone (nomophobia) pada mahasiswa jurusan ilmu komunikas fakultas ilmu sosial dan politik universitas diponegoro semarang. *Jurnal Empati*, 6(2), 15–20.

Batong, M & Uyun, Q. (2014). *Hubungan antara religiusitas dengan perilaku seksual pranikah pada mahasiswa* (Undergraduate Thesis). Fakultas Psikologi dan Ilmu Sosial Budaya UII Yogyakarta.

Buctot, D. B; Kim, N; & Kim. S. H. (2020). The role of nomophobia and smartphone addiction in the lifestyle profiles of junior and senior high school students in the Philippines. *Social Sciences & Humanities Open*, 2(1), 100035. https://doi.org/10.1016/j.ssaho.2020.100035.

Busch, P. A & McCarthy, S. (2021). Antecedents and consequences of problematic smartphone use: A systematic literature review of an emerging research area. *Computers in Human Behavior*, 114, 106414. https://doi.org/10.1016/j.chb.2020.106414.

Bragazzi, N. L. & Puente, G. D. (2014). A proposal for including nomophobia in the new DSM-V. *Psychology Research and Behavior Management*, 7, 155–160. https://doi.org/10.2147/PRBM.S41386

Bragazzi, N. L., Re, T. S., & Zerbetto, R. (2019). The relationship between nomophobia and maladaptive coping styles in a sample of italian young adults: insights and implications from a cross-sectional study. *JMIR mental health*, 6(4), e13154. https://doi.org/10.2196/13154

Carter, E. C., Mccullough, M. E., & Carver, C. S. (2012). The mediating role of monitoring in the association of religion with self-control. *Social Psychological and Personality Science*, 3(6), 691–697.

Cha, S.-S., & Seo, B.-K. (2018). Smartphone use and smartphone addiction in middle school students in Korea: Prevalence, social networking service, and game use. *Health Psychology Open*, 5(1), 205510291875504. https://doi.org/10.1177/2055102918755046

Cohen, J. (1992). A power primer. *Psychological Bulletin*, 112(1), 155–159

Davey, S., & Davey, A. (2014). Assessment of smartphone addiction in indian adolescents: a mixed method study by systematic-review and meta-analysis approach. *International Journal of Preventive Medicine*, 5(12), 1500–1511.

De Ridder, D. T. D., Lensvelt- Mulders, G., Finkenauer, C., Stok, F. M. & Baumeister, R. F. (2018). Taking stock of self- control: A meta- analysis of how trait self- control relates to a wide range of behaviors. In R. F. Baumeister. *Self-Regulation and Self-Control: Selected works of Roy F. Baumeister*. New York: Routledge.

Geyer, A. L & Baumeister, R. F. (2005). Religion, Morality, and Self-Control: Values, Virtues, and Vices. In R. F. Paloutzian & C. L. Park (Eds). *Handbook of the psychology of religion and spirituality* (pp. 412–432). New York: The Guilford Press.

Gonçalves, S., Dias, P., & Correia., A. (2020). Nomophobia and lifestyle: Smartphone use and its relationship to psychopathologies. *Computers in Human Behavior Reports*, 2, 100025.
Hanum, Z. (2021, March 7). Kemenkominfo: 89% Penduduk Indonesia Gunakan Smartphone. *Media Indonesia.*
Harish, B. R & Bharath, J. (2018). Prevalence of nomophobia among the undergraduate medical students of mandya institute of medical sciences, mandya. *Journal of Community Medicine & Public Health*, 5 (12), pp. 5455–5459, 10.18203/2394-6040.ijcmph20184833.
King & Silva. (2014). "Nomophobia": impact of cell phone use interfering with symptoms and emotions of individuals with panic disorder compared with a control group. *Clinical practice and epidemiology in mental health: CP & EMH*, 10, 28. doi: 10.2174/1745017901410010028.
Laird, R. D., Marks, L. D., & Marrero, M. D. (2011). Religiosity, self-control, and antisocial behavior: Religiosity as a promotive and protective factor. *Journal of Applied Developmental Psychology*, 32(2), 78–85. https://doi.org/10.1016/j.appdev.2010.12.003.
Lopez-Fernandez, O.; Honrubia-Serrano, L.; Freixa-Blanxart, M & Gibson, W. (2014). Prevalence of problematic mobile phone use in british adolescents. *Cyberpsychology, Behavior, and Social Networking*, 17:2, 91–98 https://www.liebertpub.com/doi/10.1089/cyber.2012.0260#
McCullough, M. E., & Willoughby, B. L. B. (2009). Religion, self-regulation, and self-control: Associations, explanations, and implications. *Psychological Bulletin*, 135(1), 69–93.
O'Dea, S. (2021, Sept 10). *Smartphones – Statistics & Facts*. Statista.
Qutishat, M.; Lazarus, E. R.; Razmy, A. M.; & Packianathan, S. (2020). University students' nomophobia prevalence, sociodemographic factors and relationship with academic performance at a University in Oman. *International Journal of Africa Nursing Sciences*, 13, 100206.
Reutter, K. K, & Bigatti, S. M. (2014). Religiosity and Spirituality as Resiliency Resources: Moderation, Mediation, or Moderated Mediation? *Journal for the Scientific Study of Religion*, 53(1):56–72.
Rahmantori, A., & Wijaya, H.E. (2018). *Hubungan antara kontrol diri dan prokrastinasi akademik pada mahasiswa.* (Undergraduate Thesis). Fakultas Psikologi dan Ilmu Sosial Budaya Universitas Islam Indonesia, Yogyakarta.
Rezki, Indriati, & Erwin (2018). Proceedings Riau International Nursing Conference (RINC) 2018. Incorporating Technology and Ethics in Advancing Nursing Education and Practice. Faculty of Nursing, Universitas Riau. 109–113. https://keperawatan.unri.ac.id/wp-content/uploads/2019/03/Proceedings- RINC2018.pdf
Saputra, Fahmi. (2018). *Hubungan antara extraversion personality dan kecenderungan nomophobia pada mahasiswa* (Undergraduate Thesis). Fakultas Psikologi dan Ilmu Sosial Budaya Universitas Islam Indonesia, Yogyakarta.
Sunday, O. J., Adesope, O. O., & Maarhuis, P. L. (2021). The effects of smartphone addiction on learning: A meta-analysis. *Computers in Human Behavior Reports*, 4, 100114.
Servidio, R. (2021). Self-control and problematic smartphone use among Italian University students: The mediating role of the fear of missing out and of smartphone use patterns. *Curr Psychol*, 40, 4101–4111. https://doi.org/10.1007/s12144-019-00373-z
Shim, J.Y. (2019). Christian Spirituality and Smartphone Addiction in Adolescents: A Comparison of High-Risk, Potential-Risk, and Normal Control Groups. *J Relig Health,* 58, 1272–1285.
Tams., Legoux & Léger. (2018). Smartphone withdrawal creates stress: A moderated mediation model of nomophobia, social threat, and phone withdrawal context. *Computers in Human Behavior*, 81, 1–9.
Tangney, J. P., Baumiester, R.F., & Boone, A.L. (2004). High self-control predicts good adjusment, less pathology, better grades, and intepersonal success. *Journal of Personality*, 72(2), 271–322.
Watterson, K., & Giesler, R. B. (2012). Religiosity and self-control: When the going gets tough, the religious get self-regulating. *Psychology of Religion and Spirituality*, 4(3), 193–205.
Yonker, J. E., Schnabelrauch, C. A., & DeHaan, L. G. (2012). The relationship between spirituality and religiosity on psychological outcomes in adolescents and emerging adults: A meta-analytic review. *Journal of Adolescence*, 35 (2), 299–314. https://doi.org/10.1016/j.adolescence.2011.08.010.
Yildirim & Correia. (2015). Exploring The Dimensions of Nomophobia: Development and Validation Of A Self-Reported Questionnaire. *Computers in Human Behavior*, 49, 130–137.

Improving Assessment and Evaluation Strategies on Online Learning – Wonorahardjo, Karmina & Habiddin (eds)

Compromising peer assessment using Google Form in an online essay-writing course

N. Ariani* & Y. Febrianti
Universitas Negeri Malang, Indonesia

ABSTRACT: This study investigates the process of development and utilization of Google Form for online peer assessment in the three essay-writing course classes in a tertiary level of education in Indonesia. Using a qualitative design, the Google Form was developed following an R&D procedure. The findings of the study showed that the successful development and piloting implementation of Google Form provides a foundation for a reliable assessment tool for writing courses. Moreover, it is useful as an interactive platform for student collaboration in peer-assessment activities. However, it is pivotal to identify issues in the early stage of development, which may occur during the implementation of peer-assessment activity to prevent technical challenges in the online teaching for writing courses. This study provides a clear and valuable idea for writing instructors who wish to create an online peer reviewing activity with Google Form and incorporate it into their teaching.

Keywords: Google Form, online essay-writing course, teaching

1 INTRODUCTION

The current pandemic has abruptly transitioned the conventional teaching to a full online class. This has caused some changes in the teaching of online writing class, especially in tertiary level of education. Therefore, writing assessment should also be adjusted to suit the online delivery of most projects and assignments on writing courses. To some extent, the alternative online assessment for students' writing needs to consider the opportunity for students to receive feedback to improve their writing.

As one of the commonly used and practiced assessments in the classroom, peer assessment has been importantly claimed to contribute to students' writing performance. Thus, teachers have considered it essential to carry out peer assessment to provide alternative assessment for the students instead of fully relying on teachers' assessment. This form of assessment allows students to play the roles of assessors and train themselves in evaluating and revising their work as if it were from a teacher's perspective (Bound 2000). In the context of higher education, the practice of peer-assessment in academic writing is believed to contribute to students' writing. Studies have shown that peer assessment plays an important role in the writing process (Dewi et al. 2019; Casas et al. 2018; Woodhouse & Wood 2020).

With various technological apps available to accommodate the online learning, teachers in Indonesia have attempted to find ways to conduct peer assessment in their online writing classes. Pradana (2020) researched the process of online peer assessment implementation using WhatsApp and found that the activity is very effective and helpful for students in learning writing and evaluating learning process. A study from Yufrizal, Sinaga, and Malinda (2021) explored how Google Classrooms provides a platform for a peer assessment and finds that it improves students' writing skills.

Nevertheless, these studies have not addressed the practice of online peer assessment in the context of higher education in Indonesia using different platform as the tool for peer assessment. Hence, this study attempts to respond to this situation by developing Google Form as an assessment tool for writing tasks using an online and accessible application. The present study will provide a

*Corresponding Author

 DOI 10.1201/9781003261346-9

clear and valuable idea for writing instructors who wish to create an online peer reviewing activity using Google Form and incorporate it into their teaching. Specifically, this study aims to answer this research questions:

1. How is Google Form developed and utilized for online peer assessment in the essay-writing course?
2. What are the challenges in developing and implementing Google Form in the essay-writing course?

2 LITERATURE REVIEW

2.1 *The benefits of peer assessment*

The literature has presented the pivotal role of writing assessment in the context of teaching and learning writing skill. White (2019) argued that writing assessments have affected every aspect of teaching writing, including how writing assignments are carried out and responses are given to students' writing by teachers. In addition, the vast expansion of technology into all aspects of writing assessment has been extensive (Comer & White 2016).

Previous research has shown that this type of assessment encourages students to take more responsibility for the learning process, assist the development of assessment skill (Seifert & Feliks 2018), and increase their involvement, independence, and assertiveness (Ellis 2001). Other benefits of peer assessment also include the ability to write critically (Woodhouse & Wood 2020), enhance self-regulated learning (Fathi & Shirazizadeh 2019), and work effectively as one of the assessment tools to measure students' writing (Dewi et al. 2019). In addition, Ng and Yu (2021) mentioned that peer assessment demonstrates the value and importance of incorporating strategies planned specifically to foster and maintain a positive attitude in the peer assessment process. These positive findings regarding peer assessment have encouraged teachers and students to practice it in the classroom to improve students' writing skills.

The practices of online assessment in the classroom have been researched and some have shown the positive impact of online peer assessment. Chen (2021) found that peer assessment in a blended learning context creates positive learning outcomes and more interesting attitudes for students. Ahmed and Al-Kadi (2021) compared the preferences for online and face-to-face peer assessment and found that online peer assessment is ineffective due to the distracting aspect of various forms of technology. Akcay et al. (2021) investigated the perception of pre-service elementary teachers about online and traditional peer assessment and found that the combination of peer and instructor evaluation and self-assessment can give a better validity and objectivity of assessment. These studies have indicated the positive role of peer assessment in supporting the teaching and learning activities during the pandemic.

Peer assessment in writing courses allow students to reflect on their own works, understand the advantages and disadvantages of their own works, integrate or retain the advantages of their works, and avoid the same mistakes or disadvantages (Topping 2009; Yu & Wu 2013). Studies conducted on L2 learners show that learners who work collaboratively develop grammatical accuracy, vocabulary, and discourse (DiCamilla & Anton 1997; Storch 2005). Thus, incorporating peer assessment in writing classes, especially in L2 contexts, accelerates the revision and editing process (Moloudi 2011). In addition, Caulk (1994) examined the effectiveness of peer assessment in academic writing and found that 89% participants made useful comments and 60% came up with good suggestions. It can be clearly seen that students' writing benefited from the practice of peer assessment.

2.2 *Online peer assessment in writing course in Indonesia*

The online practice of peer assessment has been extensive during the pandemic. However, few previous studies in Indonesian EFL context have reported how peer assessment was carried out online in writing course. A study that was conducted by Dewi et al. (2019) was done before the pandemic and was intended to discover views of peer assessment practice in graduate students' academic writing. The assessment happened when the students exchanged questions with their peers regarding their problems in the process of writing academic papers. The study presented results on the process of peer-assessment practice and the difficulties during the implementation. The students shared positive feedback on the peer-assessment practice, and students' responses indicated that they enjoyed sharing knowledge to their peers' works, including the content of writing. To minimize difficulties in doing the assessment, the lecturers started by explaining what peer assessment process was and helped the students to understand the process.

One study that was conducted to investigate the peer assessment practice in online writing class was the one by Amalia (2021). The subjects were 14 eleventh-graders of Senior High School whose English class was carried out through Google Meet and WhatsApp group and who had experience in doing peer assessment. The teacher started the process by explaining how students should make use of rubric that includes content, organization, grammar, vocabulary, and mechanics for assessing their friends' writing as well as a guideline for them to self-evaluate their own works. The teacher then sent the rubric along with the Google Form link to WhatsApp group and the students were informed about the deadline for submitting the assessment results. The Google Form assessment tool was in the form of Likert scale and paragraph to give comments. However, the students were not trained, even not clearly informed about the assessment criteria that consisted of scores and the comments. Moreover, the students only focused on revising their drafts and submitting them to the teacher without discussing the results with each other.

Pradana (2020) investigated the implementation of online peer assessment through WhatsApp during writing activities and vocational high school students' responses toward the practice. Using case study as the research design, the researcher discovered that the peer assessment comprised of four stages of implementation, namely introduction, preparation, evaluation, and application. The introduction gave students opportunity to understand the definition, purpose, and benefits of peer assessment. Then the teacher explained the assessment rubric, and the students were trained in some groups before they began the implementation. Although there were scores provided in the rubric, the teacher would only focus on students' comments on their friends' writing. During the implementation, the teacher guided the students who were providing comments on the peers' works. Finally, the process was evaluated by the teacher who then gave comments on the implementation and helped the students solve problems that they faced while assessing writing. The rubrics from each group were collected by the teacher to compare with assessment conducted previously by the teacher to see the objectivity of peer assessment

3 METHODS

3.1 *Research design*

This study adapted the R&D model by Borg & Gall (1983) with the procedure starting with 1) Need Analysis; 2) Product Development, 3) Peer Review, 4) Try-out, 5) Product Revision; 6) Reflection.

3.2 *Participants and the contexts*

The study is situated in the context of the teaching of an essay-writing course in an English department at Universitas Negeri Malang, Indonesia. The participants were 50 students in the Department of English who are required to learn standard five-paragraph essays comprising example essay, comparison/contrast essay, classification essay, process analysis essay, and cause and effect essay from the essay-writing course. There are two writing lecturers with 50 students participating in the study.

3.3 *Data analysis procedure*

In the product development stage, Google Form for the essay-writing assessment is created in accordance with the guidance given in the textbook of essay writing (Smalley et al. 2001). The Google Form assessment is created to guide students' self-assessment activity for each type of essay, including the points of checking for the number of the paragraph, the transition signals used in the essay, and the validity of the thesis statement. For the scoring purpose, students are guided to use a system in which they can rate their essay using a scale from 10 to 100. The Google Form is also designed for use as an instrument for peer review. For this purpose, a different mechanism is required. The peer review process of the product was carried out by having the department writing instructors to review the product and then provide a comment and feedback. Once the peer review was completed, the Google Form was administered to the three essay-writing classes. The reflection stage was carried out after the essay-writing course was completed at the end of the semester.

3.4 *Data collection and analysis*

Data collection was carried out by administering the Google Form three essay-writing course with 50 students in the implementing stage. The responses from the Google Form were taken as samples of responses in providing feedback. The information from the responses is also useful to gain insights into the reflection about the challenges and implication from the overall study.

4 FINDINGS AND DISCUSSION

4.1 *The development of the Google Form*

During the planning stage, the lecturers involved in the teaching of essay writing discussed the material as the main reference throughout the writing course. Students in this course are required to learn and write five different types of essays. Hence, there were five discussion rounds. At each discussion, students would initially learn the characteristics of each essay type. In particular, students' writing will be assessed at the end of the course.

This learning method was chosen as the study follows Smalley et al. (2001), wherein a "classic" book was used as a guideline and main material for developing the lesson plans for the essay-writing course. The handbook by Smalley et al. (2001) especially fits the assessment purpose because: 1) it has a clear description and prescription of various types of essays that can guide novice students during essay development; and 2) the book's description of each essay type is easy to follow, especially for the context of English as a second language. Although the book itself does not prescribe a particular assessment the description and prescription written specifically for each essay support the development of assessment points in the Google Form. Thus, it makes the content of this book fit for developing the assessment using Google Form.

Discussion and exercises on the required linguistics features for the essays then follow. Smalley et al.'s study (2001) is used as a guideline and main material for developing lesson plans for the essay-writing course based on a few considerations. First, it has a clear description and prescription of various types of essays as a guideline for novice students to develop the essays. Second, the description in the book for each essay type is easy to follow, especially for the context of English as a second language. The book itself does not prescribe a particular assessment tool. Therefore, the description and prescription of each essay are adopted to develop the points of assessment that will be used in the Google Form.

During the design stage, a number of considerations were measured in creating the assessment tool using an online application. Initial need analysis for the course was conducted in a regular meeting before the beginning of semester to evaluate the preparation for an online delivery of the course. Here, the talk regarding the teaching platform, instructional materials, and assessment were discussed to reach the same perspectives regarding the new situation of online teaching of writing course. It was quite common that one lecturer handles more than one classes of essay writing. With large number of students, it becomes even more necessary to create a practical and time-efficient tool to accommodate the change in the delivery of the materials for the students. The students of this particular cohort are digital natives who are tech-savvy. They are familiar with the use of technology and computer applications. This brings optimism in the design of an assessment tool using an online application.

It should be noted, however, while Google Form has provided the platform, the content of the assessment should be made based on robust principles upon which the essay-writing course is developed. In this way, the development of the tool would ensure the legitimacy, reliability, and validity of the Google Form as an assessment tool. In this case, the guideline is taken from the handbook used for the class (Smalley et al. 2001).

The development stage of the assessment tool as the third stage deals with the practical implementation of what has been outlined in the design stage. Considering that the teaching and learning materials were developed using a variety of digital materials, the Google Form assessment tool is prepared as an evaluation, the points of which are adjusted to the types of essays. In this study, the observed essay-writing course is the four-credit course with 32 meetings. There are five essay types, which need to be written by the students for the entire semester. As a pilot project, the Google Form is made to comply with the generic of the essay types. Therefore, Google Form assessment is created following the guidance of Smalley et al. (2001), covering aspects of the characteristics of a three-part essay, thesis statement, introductory paragraph, developmental paragraph, concluding paragraph, and transition between paragraphs.

These points are then interpreted into the Google Form system. The technical development of the Google Form assessment adopted essay guidelines from the handbook as the evaluation points. The guidelines are transformed into statements divided into the categories of: organization, content, grammar, vocabulary, and mechanics. These categories are considered as they are the aspects of writing that are essential both in writing and for students to pay attention. In each of the categories, the statements are made in accordance with the evaluation points from Smalley (2001). In every point, students are guided to use a Likert scale in which they can rate the essay using an ascending scale from 1 to 5 to indicate the lowest to the highest score, respectively. At the end of the Google Form, students are asked to write about the general comment of the essay that they read. This open-ended part of assessment is required for students to fill in based on their analysis of other students' writing in a critical and constructive way (Luxton-Reilly 2009). By having additional comment instead of only the Likert scale scoring, students can be more benefitted as it is possible for each of them to find which part of writing that needs some changes and how they are supposed to revise it.

4.2 *The implementation of the Google Form*

The next stage is the try-out. At this stage, the designed Google Form assessment is piloted to the class. The overall intention to create the Google Form assessment tool is to facilitate the peer assessment for the students in the class. Peer assessment is the final stage of every cycle in the writing of each essay type. As part of the curriculum, peer assessment can encourage students to take more responsibility for the learning process and assist the development of assessment skills (Seifert & Feliks 2018). It also increases their involvement, independence, and assertiveness (Ellis 2001). Generally, the development of the assessment tool using Google Form is designed in such a way that students can access and operate from any available digital device. Note that the assessment form can also be used independently as a form of self-assessment. A few challenges, however, were found in the piloting process.

The objective of the try-out is to know whether students could understand the procedure in using the Google Form for self- and peer-assessment purposes. One of the first challenges in the development is transforming the guideline from Smalley et al. (2001) into statements that will be used in the Google Form. The wordings and phrasing should be made clear enough so that the students understand the statement. This is important to lead the students to identify each point of evaluation in the essay that they read. The Google Form should also consider clear instructions for easy understanding for independent use for students.

The utilization of online tools for peer assessment can serve significant benefits for the management of the assessment process (Ballantyne et al. in Luxton-Reilly 2009) so that the practice can be more efficient. Having a particular online tool for peer assessment during this pandemic enables students to administer students' works to be reviewed by others as well as to submit the result of peer assessment to the lecturers. Thus, despite the limitation of not having regular peer assessment at physical classroom, the availability of sufficient assessment tool that is designed to help students review their friends' essays is believed to support the improvement of online assessment practice.

The decision to utilize Google Form is due to students' familiarity toward the platform that is commonly used both inside and outside the context of university courses. It is also based on a recommendation from Technology Acceptance Model (in Kaewsaiha & Chanchalor 2020) that the tool used for peer assessment should be something that is familiar to the users, relevant to the purpose of the assessment, as well as can be applied with existing resources.

As part of the piloting project, the Google Form administration recorded students' responses for the last question in the form, i.e., an open-ended question requiring students to write their comments on their peers' essay. The responses for this question reflect the students' understanding on the elements of a good essay. Following are a few samples of responses from students:

Excerpt 1:

Overall, the content of the essay is great. The introduction is good enough to trigger the reader's curiosity. However, some sentences actually can be written in a more effective way. I also noticed that the use of the word "sim" is actually unclear for those who don't play The Sims. Perhaps it's better to add a brief definition of the word.

Excerpt 2:

Overall, the essay is good, but I think there are several things that can be improved. When I check the essay, I think there are several grammatical errors and some word choices that are not capable enough. The concluding sentences have to be clearer.

Excerpt 3:

The overall content has been done well. Unfortunately, it was not written in a five-paragraph-essay, so the thesis is only developed into two developmental paragraphs. The author also uses the transition devices well although I still find several grammar mistakes (the use of present tense is wrong, by having to be but not followed by v-ing). Even so, the grammar mistakes do not really disturb my experience in reading it, though it would be better if the author could write it with no errors in grammar.

Excerpt 4:

I think A's essay is good. The introductory paragraph is interesting. For developmental paragraphs, it also contains topic sentences that are relevant with supporting sentences and thesis statements. However, there are several typos and the transitions between paragraphs must be improved a little bit. For the conclusion, the paragraph restates what is in the thesis statement and does not introduce any new topics to be discussed.

Each of the excerpts commented on the essays based on the elements of the essay that they found during the reading. This implies that students understood what is expected of them when they used the Google Form for the assessment. Moreover, students' responses showed that they expressed understanding of the writing aspects that they found in their friends' works as the peer assessment is mentioned by Kaufman and Schunn (2010) to be a cognitively demanding activity that helps students grow their knowledge of the subject matter. Beyond this, it also implies that the Google Form has been developed using clear instruction to help users, i.e., the students for a practical use of the Google Form as an assessment tool. This also confirms previous study on the use of Google applications for peer assessment that had been done previously (Kaewsaiha & Chanchalor 2020) where students' attitudes toward the implementation of the online assessment were positive. This included students' views on convenience of the tool, understandable instructions, and understandable grading criteria.

5 CONCLUSION

With the ministry's plan to continue teaching and learning activities online, either in the pandemic situation or the future independent learning program, online teaching and learning activities need to be prepared more optimally, both the materials and the assessment system. This study has offered a basic development of a practical essay-writing assessment using Google Form. Efficacy of the tool offered in this study lies in its familiarity and the practicality to use with the current cohorts of students who are digital natives. Students' comments on their peers' works also indicate that the instructions can be understood as each of them shows how they can provide their opinions clearly regarding writing aspects that were produced by other students.

Further investigation, however, needs to be conducted in a number of aspects. The tool should be tried out in other writing courses more extensively and intensively to gain better insights into its applicability, validity, and reliability. More understanding about the use of the tool can be obtained by gathering students' perception of its application in the essay-writing course. This can cover students' perceptions of the instructions, grading criteria, their confidence when using the tool, and whether the tool helps them in improving motivation better rather than only feedback and grade from the lecturers. Prior to, during, and after the trials, focus group discussions with the teachers and students can be conducted to revise, improve, and re-develop the tool as a robust assessment instrument.

As a final note, the study considers that essay-writing courses in different contexts may be designed with different principles and prepared with different handbooks. This study is a good reference for developing Google Form that may be suited to assessment tools for essay-writing courses in different contexts. Replicability of Google Form assessment is offered in this study, hence, it is possible to use it in different contexts of essay-writing class using different handbooks.

REFERENCES

Ahmed, R., & Al-Kadi, A. 2021. Online and face-to-face peer review in academic writing: Frequency and preferences. *Eurasian Journal of Applied Linguistics* 7(1): 169–201.

Akcay, A. O., Guven, U., & Karahan, E. 2021. The Views of Pre-Service Elementary Teachers About Online and Traditional Peer Assessment. *International Journal of Assessment Tools in Education* 8(2): 409–422.

Amalia, N. R. 2021. The Implementation of Peer Assessment in Online Writing Class at Senior High School in Bawean. *Research on English Language Teaching in Indonesia* 9(1): 179–186.

Borg, W. R., & Gall, M. D. 1983. *Educational research: an introduction*. New York: Longman.

Casas, M. R., Nuno, N., Pons, F. & Cunillera, T. 2019. The difference impact of a structured peer-assessment task in relation to university undergraduates' initial writing skills. *Assessment & Evaluation in Higher Education* 44(5): 653–663.

Caulk, N. 1994. Comparing Teacher and Student Responses to Written Work. *Teachers of English to Speakers of Other Languages, Inc.* 28(1): 181–188.

Chen, J. 2021. The Effectiveness of Peer Assessment in EFL Blended Learning Environments. *2nd International Conference on Computers, Information Processing and Advanced Education.*

Corner, D., & White, E. M. 2016. Adventuring into MOOC Writing Assessment: Challenges, Results, and Possibilities. *College Composition and Communication* 67: 318.

Dewi, E. W., Nurkamto, J. & Drajati, N. A. 2019. Exploring peer-assessment practice in graduate students' academic writing. *A Journal on Language and Language Teaching* 22(1).

DiCamilla, F. J., & Anton, M. 1997. Repetition in the collaborative discourse of L2 learners: A Vygotskian perspective. *Canadian Modern Language Review* 53(4): 609–633.

Ellis, G. 2001. Looking at Ourselves—Self-Assessment and Peer Assessment: Practice examples from New Zealand. *Reflective Practice* 2(3): 289–302.

Fathi, J., & Shirazizadeh, M. 2019. Fostering Self-Regulated Learning of Iranian EFL Students: An Investigation of the Effect of Self and Peer Assessment in L2 Writing. *Journal of Foreign Language Research* 9(1): 123–146.

Kaewsaiha, P., & Chanchalor, S. 2020. The Development of an Online Peer Assessment Tool Using Google Applications. *ICDTE 2020.*

Kaufman, J. H., & Schunn, C. D. 2010. Students' perceptions about peer assessment for writing: their origin and impact on revision work. *Instructional Science* 39(3): 387–406.

Luxton-Reilly, A. 2009. A systematic review of tools that support peer assessment. *Computer Science Education* 19(4): 209–232.

Moloudi, M. 2011. Online and Face-to-face Peer Review: Measures of Implementation ESL Writing Classes. *The Asian EFL Journal.*

Ng, W. S. & Yu, G. 2021. Students' attitude to peer assessment process: A critical factor for success. *Interactive Learning Environments.*

Seifert, T., & Feliks, O. 2018. Online self-assessment and peer-assessment as a tool to enhance student-teachers' assessment skills. *Assessment & Evaluation in Higher Education.*

Smalley, R. L., Ruetten, M. K., & Kozyrev, J. R. 2001. *Refining Composition Skills: Rhetoric and Grammar, 5th Edition*. Boston: Heinle & Heinle Thomson Learning.

Storch, N. 2005. Collaborative writing: Product, process, and students' reflections. *Journal of Second Language Writing* 14(3): 153–173.

Topping, K. J. 2009. Peer Assessment. *Theory into Practice* 48(1): 20–27.

White, E. 2019. Revisiting twenty-five years of writing assement. *Assessing Writing*. 42.

Woodhouse, J., & Wood, P. 2020. Creating dialogic spaces: developing doctoral students' critical writing skills through peer assessment and review. *Studies in Higher Education.*

Yufrizal, H., Sinaga, T., & Malinda, R. P. 2021. A Quasi Experimental Study on google classroom mediated peer assessment of English subject at high school in Indonesia. *International Journal of Research and Innovation in Social.*

Yu, F., & Wu, C. 2013. Predictive Effects of Online Peer Feedback Types on Performance Quality. *J. Educ. Technol. Soc.* 16: 332–341.

Improving Assessment and Evaluation Strategies on Online Learning – Wonorahardjo, Karmina & Habiddin (eds)
© 2022 copyright the Authors, ISBN: 9781032198903
Open Access: www.taylorfrancis.com, CC BY-NC-ND 4.0 license

Student self-regulated learning strategies during Covid-19 pandemic

N. Eva*
Universitas Negeri Malang, Indonesia

S. Andayani
Universitas Airlangga, Indonesia

ABSTRACT: This study aims to map the Self-Regulation Learning (SRL) strategy used by students during the Covid-19 pandemic. This research was conducted using a written interview with open questions. As a result, the strategies that were widely used by students during distance learning during the Covid-19 pandemic were seeking peer assistance, environmental structuring, reharsing and memorizing, seeking information, and reviewing text. The strategies used by individuals are not only of one form, they use more than one strategy in learning regulation. Some strategies are also carried out only in certain contexts.

Keywords: self-regulated learning, Covid-19 pandemic

1 INTRODUCTION

The Covid-19 pandemic, which began in 2019, brought major changes for humans. The call for quarantine to minimize transmission is something that cannot be avoided by everyone. Everyone is required to limit all physical activities to meet other people. This restriction directly leads to the use of the internet network to connect with other people. Internet networks make it easier for workers to work from home. Likewise with students, they can still do learning activities at home.

This causes changes in habits in all fields, including in the field of education. The world of education undergoes major changes in the learning model. These changes are experienced from early childhood education to higher education. The learning model that was originally dominated by the offline model has turned into distance learning as a whole. So far, the distance learning model supported by online technology has the principle of "anywhere" and "anytime" (Zounek & Sudický 2013). This is an advantage of the online learning model so that students can have the freedom to acquire knowledge.

But unfortunately, distance learning also has shortcomings, such as the availability of facilities such as the latest technology and internet connectivity (Zounek & Sudický 2013). As many as 12.6% of Indonesians who do not use the internet admit to having no knowledge of the use of this technology (APJII). This happens in Indonesia, technology and internet connectivity are not evenly distributed (APJII 2018). In addition, from the individual point of view of their learners, they often find it difficult to utilize various time management, presentation, word processing, collaboration, and other types of tools for personal learning purposes; ineffective way of implementing technology and even complete rejection of any electronic learning solution (Zounek & Sudický 2013).

At the higher education level, the use of online learning models has been widely used and collaborated with offline learning before the Covid-19 pandemic, the difference is that during the Covid-19 period, learning and guidance was completely online. Direct social contact is also limited, so the learning material presented is not comprehensive, especially for materials that are practical.

*Corresponding Author

A change in the learning model can affect student Self-Regulation Learning (SRL) (Viberg et al. 2020). Good SRL skills will determine achievement or learning outcomes (Ejubovic & Puška 2019).

In the past 5 years, research on SRL has increased, mostly examining how much a person's SRL ability level and its correlation or influence on other variables or vice versa. So far, there have been many studies on SRL that measure the level of individual SRL's ability to produce high to low score categories continuously (Callan 2014). In fact, regardless of the level of SRL, each individual has their own strategy in doing SRL (Zimmerman & Pons 1986). The research conducted shows that this strategy is not limited to individuals who have high SRL abilities (Zimmerman & Martinez-Pons 1990). A good and appropriate SRL strategy will affect learning outcomes (Effeney et al. 2013). Several studies have not examined what SRL strategy is used by individuals. Therefore, this study tries to map what SRL strategy students used during the Covid-19 pandemic.

2 METHODS

Research on student SRL strategies during the Covid-19 pandemic was carried out with the aim of understanding what SRL strategies were used, how often they were carried out, and the context in which they were used. The sample was selected using random sampling method. The research subjects in this study were university students in Indonesia. The total subject is 50 people.

Data was collected by survey method containing open-ended questions. The questions given in the survey use a guided question called the Self-Regulated Learning Interview Schedule (Zimmerman & Martinez-Pons 1990). There are eight open-ended questions and respondents are asked to share their learning experiences in writing in different contexts, in online classroom situations, at home, when completing writing assignments outside class, assignments outside class, when preparing for and taking tests, and when poorly motivated. Then the researchers analyzed thematically, what strategies were used by students. Themes were categorized based on SRL theory (Zimmerman & Martinez-Pons 1990) into 15 strategies.

3 RESULT AND DISCUSSION

3.1 *Result*

The results of the written interviews showed that five SRL strategies that were most widely used by students during the Covid-19 pandemic were seeking peer assistance, environment structuring,

Table 1. College student SRL strategies.

Strategy	Total	Percentage
Seeking peer assistance	49	98
Environment structuring	48	96
Rehearsing and memorizing	48	96
Seeking information	47	94
Reviewing text	46	92
Organizing and transforming	45	92
Self-Consequating	45	90
Reviewing notes	45	90
Keeping records and monitoring	42	84
Other	40	80
Seeking teacher assistance	35	70
Self-evaluation	34	68
Goal setting and planning	27	54
Seeking adult assistance	25	50
Reviewing test	8	16

reharsing and memorizing, seeking information, and reviewing text. Through these results, it is known that each respondent uses more than one strategy when implementing learning regulations. Meanwhile, the three strategies that are least used are reviewing tests, seeking adult assistance, and goal setting and planning. The reviewing test strategy is the only strategy where the users are below 50%.

3.2 *Discussion*

The data presentation above shows the five SRL strategies with the most users, namely seeking peer assistance, environment structuring, rehearsing and memorizing, seeking information, and reviewing text. First, the strategy of seeking peer assistance is indicating student-initiated efforts to solicit help from peers. This strategy ranks first. This strategy takes advantage of social influences and human functions to help individuals conduct learning regulation. However, during a pandemic, individual interactions are physically limited so that people try to meet their social needs through any media (William 2020). The need will be higher in intensity if not immediately met the need will be higher in intensity if not immediately fulfill (Weston et al. 1978). Therefore, activities that involve interaction with other people online are getting bigger. This can also be influenced by the characteristics of Indonesian society that are inclined toward a collectivism culture (Mankowski & Thomas 2000), so indirectly the desire to affiliate and socialize with other people is getting higher. Basically, humans are social creatures, the desire to interact with other people either directly or not is an important need (Frank et al. 2015).

The second strategy is environment structuring. Environment structuring indicated student-initiated efforts to select or arrange the physical setting to make learning easier. Environment structure is not only related to the conditions of the room when studying, including managing personal physical conditions (Puteh et al. 2015). For example, some people can learn well if it is accompanied by listening to music, or they have to clean themselves first, some even feel that they can learn well by doing worship first. During the Covid-19 pandemic, clean living habits increased (Toussaint et al. 2020). Some people who previously did not really keep their environment clean can turn into people who are very aware of the cleanliness of their surroundings. This is all done solely to provide a sense of security and comfort when doing activities. In fact, several studies have shown a strong relationship between physical structure and student learning comfort (King & Marans 2013; Puteh et al. 2015). This also applies to learning using computer based as it is today (Yildirim et al. 2011).

The next strategy is rehearsing and memorizing. This strategy indicated student-initiated efforts to memorize material by overt or covert practice. The results of the interviews obtained provide information that the student's effort to remember the material that must be studied is to rewrite the material that has been obtained during lectures. They admit that by rewriting the material to be studied, it will help them remember the material. There has been a lot of research on the relationship between writing and memory and the results prove that writing improves long-term memory (Linderholm & Abrams 2012). Writing will also improve brain development (James & Engelhardt 2012). The results of the interview also indicated that the notes that the students made back up were mostly written by hand, not typed. A study comparing the effect of writing using a pen and a keyboard, the results show that writing by hand will increase literacy skills more than a keyboard (Berninger et al. 2009).

The next strategy is seeking information. Seeking information indicated the student-initiated efforts to secure further task information from non-social sources when undertaking an assignment (Zimmerman & Martinez-Pons 1990). The results of the interview showed that students used the internet to help them complete their college assignments and even studied and searched for literature or reference books via the internet. During the Covid-19 pandemic, the internet was used to help distance learning so that the intensity of students looking for information via the internet increased (Viberg et al. 2020). In addition, the internet makes it easy for students to find specific information according to their needs.

Apart from the five strategies above, there are still many other strategies used by students in regulating their learning regulations. Three strategies are at the bottom of the list, namely, text review, seeking adult assistance, and goal setting and planning. The review test strategy is only used when students are completing assignments or taking exams. Students rarely use seeking adult assistance strategies, they prefer to use seeking peer assistance and seeking teacher assistance. The peer assistance strategy further facilitates the student's need to stay connected and interact with their social environment in the online learning system (Tenenbaum et al. 2019). In addition, peer support can improve learning engagement and academic achievement (Ansari & Khan 2020; Lai et al. 2019).

Goal setting and planning strategies aim for someone to have goals and strategic plans to achieve them. This strategy helps students plan to complete lectures well (Yusuff 2018). In fact, this strategy is in the bottom three. This can be caused by changes in individuals because priorities during the Covid-19 pandemic will be different from those before the pandemic. During a pandemic, most of the priority of one's goals is health and survival (Khetrapal & Bhatia 2020). Therefore, not all strategies are carried out by students in their learning regulations. Some strategies are only used in certain contexts.

4 CONCLUSION

During the Covid-19 pandemic, students organized themselves in learning using several strategies collaboratively. There are five strategies that are most often used in regulating their learning regulations, namely seeking peer assistance, environmental structuring, rehearsing and memorizing, seeking information, and reviewing text. Students do not use one form of strategy in regulating their learning regulations. They collaborate one form of strategy with another. In addition, not all strategies are suitable for students to use when carrying out learning activities. Some strategies are only used in certain activities.

REFERENCES

Ansari, J. A. N., & Khan, N. A. (2020). Exploring the role of social media in collaborative learning the new domain of learning. *Smart Learning Environments*, *7*(1), 9. https://doi.org/10.1186/s40561-020-00118-7

Beeson, S. A. (1996). The effect of writing after reading on college nursing students' factual knowledge and synthesis of knowledge. *The Journal of Nursing Education*, *35*(6), 258–263.

Berninger, V. W., Abbott, R. D., Augsburger, A., & Garcia, N. (2009). Comparison of Pen and Keyboard Transcription Modes in Children with and without Learning Disabilities: *Learning Disability Quarterly*. https://doi.org/10.2307/27740364

Callan, G. L. (n.d.). *Self-regulated Learning (SRL) Microanalysis for Mathematical Problem Solving: A Comparison of a SRL Event Measure, Questionnaires, and a Teacher Rating Scale*. 196.

Effeney, G., Carroll, A., & Bahr, N. (n.d.). *Self-Regulated Learning: Key strategies and their sources in a sample of adolescent males*. 17.

Ejubovic, A., & Puška, A. (2019). Impact of self-regulated learning on academic performance and satisfaction of students in the online environment. *Knowledge Management and E-Learning*, *11*, 345–363. https://doi.org/10.34105/j.kmel.2019.11.018

Frank, B., Enkawa, T., & Schvaneveldt, S. J. (2015). The role of individualism vs. collectivism in the formation of repurchase intent: A cross-industry comparison of the effects of cultural and personal values. *Journal of Economic Psychology*, *51*, 261–278. https://doi.org/10.1016/j.joep.2015.08.008

James, K. H., & Engelhardt, L. (2012). The effects of handwriting experience on functional brain development in pre-literate children. *Trends in Neuroscience and Education*, *1*(1), 32–42. https://doi.org/10.1016/j.tine.2012.08.001

Khetrapal, S., & Bhatia, R. (2020). Impact of COVID-19 pandemic on health system & Sustainable Development Goal 3. *The Indian Journal of Medical Research*, *151*(5), 395. https://doi.org/10.4103/ijmr.IJMR_1920_20

King, J., & Marans, R. (2013). *The Physical environment and the learning process: A survey of recent research.*

Lai, C.-H., Lin, H.-W., Lin, R.-M., & Tho, P. (2019). Effect of Peer Interaction among Online Learning Community on Learning Engagement and Achievement. *International Journal of Distance Education Technologies*, *17*, 66–77. https://doi.org/10.4018/IJDET.2019010105

Linderholm, T., & Abrams, L. (2012). The benefits of expressive writing onlong-term memory performance. *Long-Term Memory: Mechanisms, Types and Disorders*, 131–146.

Mankowski, E. S., & Thomas, E. (2000). The relationship between personal and collective identity: A narrative analysis of a campus ministry community. *Journal of Community Psychology*, *28*(5), 517–528. https://doi.org/10.1002/1520-6629(200009)28:5<517::AID-JCOP5>3.0.CO;2-A

Public perceptions and experiences of social distancing and social isolation during the COVID-19 pandemic: A UK-based focus group study | medRxiv. (n.d.). Retrieved October 28, 2020, from https://www.medrxiv.org/content/10.1101/2020.04.10.20061267v1

Puteh, M., Che Ahmad, C. N., Noh, N., Adnan, M., & Ibrahim, M. H. (2015). The Classroom Physical Environment and Its Relation to Teaching and Learning Comfort Level. *International Journal of Social Science and Humanity*, *5*, 237–240. https://doi.org/10.7763/IJSSH.2015.V5.460

Tenenbaum, H., Leman, P., winstone, naomi, & avery, rebecca. (2019). How Effective is Peer Interaction in Facilitating Learning? A Meta-Analysis. *Journal of Educational Psychology*.

Toussaint, L. L., Cheadle, A. D., Fox, J., & Williams, D. R. (2020). Clean and Contain: Initial Development of a Measure of Infection Prevention Behaviors During the COVID-19 Pandemic. *Annals of Behavioral Medicine: A Publication of the Society of Behavioral Medicine*, *54*(9), 619–625. https://doi.org/10.1093/abm/kaaa064

Viberg, O., Khalil, M., & Baars, M. (2020, March 23). *Self-Regulated Learning and Learning Analytics in Online Learning Environments: A Review of Empirical Research*. https://doi.org/10.1145/3375462.3375483

Weston, B. H., Fishlow, A., Schachter, O., Zamora, S., & Cristensen, C. (1978). Basic Human Needs: The International Law Connection. *Proceedings of the Annual Meeting (American Society of International Law)*, *72*, 224–240.

Yildirim, K., Capanoglu, A., & Cagatay, K. (2011). The Effects of Physical Environmental Factors on Students' Perceptions in Computer Classrooms: *Indoor and Built Environment*. https://doi.org/10.1177/1420326X11411135

Yusuff, K. B. (2018). Does personalized goal setting and study planning improve academic performance and perception of learning experience in a developing setting? *Journal of Taibah University Medical Sciences*, *13*(3), 232–237. https://doi.org/10.1016/j.jtumed.2018.02.001

Zimet, G. D., Dahlem, N. W., Zimet, S. G., & Farley, G. K. (1988). The Multidimensional Scale of Perceived Social Support. *Journal of Personality Assessment*, *52*(1), 30–41. https://doi.org/10.1207/s15327752jpa5201_2

Zimmerman, B. J. (2002). Becoming a Self-Regulated Learner: An Overview. *Theory Into Practice*, *41*(2), 64–70. https://doi.org/10.1207/s15430421tip4102_2

Zimmerman, B. J., & Martinez-Pons, M. (1990). Student differences in self-regulated learning: Relating grade, sex, and giftedness to self-efficacy and strategy use. *Journal of Educational Psychology*, *82*(1), 51–59. https://doi.org/10.1037/0022-0663.82.1.51

Zimmerman, B. J., & Pons, M. M. (1986). Development of a Structured Interview for Assessing Student Use of Self-Regulated Learning Strategies. *American Educational Research Journal*, *23*(4), 614–628. https://doi.org/10.3102/00028312023004614

Zounek, J., & Sudický, P. (2013). Heads in the cloud: Pros and cons of online learning. *International Conference Disco 2013: New Technologies and Media Literacy Education, December*, 58–63.

Improving Assessment and Evaluation Strategies on Online Learning – Wonorahardjo, Karmina & Habiddin (eds)
© 2022 copyright the Authors, ISBN: 9781032198903
Open Access: www.taylorfrancis.com, CC BY-NC-ND 4.0 license

Experience using a five-component blended learning strategy during the Covid-19 pandemic

N.C.E. Setiawan*
School of Education, Universiti Teknologi Malaysia, Malaysia
Faculty of Mathematics and Science, Chemistry Departement, Universitas Negeri Malang, Indonesia

M.S. Rosli
School of Education, Universiti Teknologi Malaysia, Malaysia

ABSTRACT: The Covid-19 pandemic has led all learning in Indonesia to be carried out online. This research aims to explore the students' experience in five-component blended learning strategy and to determine students' learning achievement in chemistry laboratory management. The blended learning approach consists of five stages, namely Discover, Learn, Practice, Collaborate, and Assess (DLPCA). The asynchronous session of teaching is done with video recordings provided in the learning management system (SIPEJAR) to allow students to learn at their own rate. The synchronous part is done using the Google Meet video conferencing platform. The course participants consisted of 41 6th semester students who took the School Chemistry Laboratory Management course at the Department of Chemistry of a public university in East Java. The participants come from seven provinces in Indonesia, spread from the western part of Indonesia to the central part of Indonesia. The data were obtained through questionnaires, observations, and portfolios. Two indicators applied for the analysis of teaching and learning experience were the students' learning achievement, as well as experience. There were two challenges faced by the students, namely the internet connection and hands-on learning experience in understanding laboratory equipment. Therefore, lecturers should find ways to increase student interaction and sustain students' interest and participation during online classes. The questionnaire also showed that most of the students were satisfied with the DLCPA strategy. Therefore, this strategy is a practicable and effective alternative adapted to complete online instruction for other undergraduate chemistry programs.

Keywords: Blended learning, Coronavirus, DLPCA

1 INTRODUCTION

The Covid-19 pandemic has impacted various levels of society all over the world (Dai & Zhao 2020; Gunawan et al. 2020). People are suffering from this outbreak (Backer et al. 2020; Chertoff et al. 2020). Due to the coronavirus outbreak, global social lifestyles have changed a lot. Although the geographical distance between countries has shortened due to the spread of the coronavirus, the social space between humans must be limited to prevent being infected (Kampf et al. 2020). Many colleges and schools have been closed due to the pandemic. Lectures are transitioned to online mode to continue learning progress for students (Marinoni et al. 2020). The shift of all teaching and learning activities using virtual platforms has presented challenges in terms of time, effort, and cost compared to traditional learning or direct instruction in class (Boyle et al. 2003).

In Indonesia, the instruction of chemistry is assigned to make students grow sufficient competence in the field of chemistry by teaching them not only necessary knowledge and skills, but also developing a positive attitude related to chemistry, which is needed when dealing with

*Corresponding Author

 DOI 10.1201/9781003261346-11

unaccustomed situations (Ministry of Research, Technology and Higher Education of the Republic of Indonesia 2015). In chemistry instruction, students must engage in authentic experiences to discover scientific facts. Furthermore, advances in technology have broadened the aim of teaching chemistry to develop students' understanding of chemical concepts and teach them how to use knowledge in solving real-life problems (Aksela, 2005; Inoue, 2010). Thus, students must collect relevant information to analyze and interpret to find the intended solution by applying a good scientific process. Modern chemistry teaching also requires students to be more enthusiastically involved in the process of teaching and learning, rather than passively accepting knowledge from lecturers (Henry 2017). As a result, students are expected to master a wide range of advanced chemical skills and thereby help make a positive contribution to global society.

Due to the uncertain future with the coronavirus outbreak and the threat of lockdowns, many instructors should change their teaching into online instruction, which could be conducted in one of three pedagogical approaches: (1) synchronous, (2) asynchronous, and (3) mixed learning strategies. In synchronous (real-time) online lectures, the instructor and students do online meetings by using video conferencing software during specified class hours, and the instructor delivers lectures on the course. Students participate in studies and can ask questions in person by asking or via live text chat. In asynchronous classes, instructors record lecture videos and upload them on a learning management system (LMS), so that students can access at the most suitable time (Alshammari et al. 2016).

A learning strategy using blended online is considered the most practical way to adapt because it brings together the benefits of synchronous and asynchronous techniques. The primary objective in using a mixed strategy is to increase student involvement in the learning process during synchronous learning. This approach is based on cognitive load theory, because novice learners find it challenging to learn new concepts (Darabi & Jin 2013; Seery 2013). This active learning pedagogy is called the "reverse classroom" approach (Olakanmi 2017). In this learning approach, traditional lectures and homework are replaced by structured tasks, such as watching instructional videos or case study videos. Asynchronous class time is devoted to problem-solving, collaborative activities, and comprehensive discussions (Pienta 2016). However, synchronous online classroom parts replace face-to-face classes to engage students with guided problem-solving activities and discussions in virtual classrooms.

One of the LMS developed by one of the universities in Indonesia is SIPEJAR. According to Surahman et al. (2019), learning using SIPEJAR can train creative thinking, collaboration, project management, and challenge. This LMS also provides various complete facilities in managing online learning (Ashar et al. 2020), such as easy access, attractive displays, discussion forums, videos, libraries, live chat, assessments, quizzes, and video conferences. This platform can also be accessed using a smartphone so that students can access it easily. With LMS, it is hoped that students can learn quickly and learn without haste because students can access learning videos and modules at a suitable time (Rosli et al. 2015).

2 METHODS

This is descriptive research using qualitative methods. All learning is carried out online using the university learning management system (LMS) called SIPEJAR. In the LMS, learning videos, learning modules, virtual discussion forums, virtual meeting links, and quizzes have been set. All student activities are fully documented in the LMS . One of the essential things from the DLPCA strategy given to students before attending the synchronous session is the learning video (Lapitan et al. 2021; Meltem 2015; Prilop et al. 2021). This learning video is made simple, easy to read, visually attractive, clear audio, and easy for students to understand.

The study was conducted between February and May 2021 at a public university in East Java, Indonesia. A total of 41 undergraduate chemistry education students who programmed the school chemistry laboratory management from two different classes in their sixth semester participated in the study. The first author of this paper taught all classes. The data collected includes questionnaires, interviews, and learning achievement. At the end of the semester, the students voluntarily fill out a questionnaire about their opinion on the course. There were 41 students answering the questionnaire. The questionnaire consisted of nine five-scale Likert items, from strongly agree to strongly disagree. The focus of the questions was on participants' views of their interests during the lecture and the quality of the instructional videos.

Interviews were purposely composed to complete the findings in the questionnaire. The participants of the interview were chosen to represent variation in responses from the questionnaire (Khery et al. 2020). The interview was 15–20 minutes long. It was recorded, transcribed, and analyzed qualitatively (Nida et al. 2021). All excerpts from interviews are translated from Indonesian. Data is provided by the participants voluntarily and is handled anonymously. The data collection process and handling fulfils the legal regulations for ethical empirical research with humans in the relevant department. The dean of the faculty grants permission to use data.

3 RESULT AND DISCUSSION

Some factors in designing an appropriate teaching approach for a school chemistry laboratory management course should be taken into account. One way is to evaluate the appropriate pedagogical model. Among learning theories, cognitivism and constructivism are considered the best applied in online classroom learning. On the one hand, cognitivism illustrates that students manage the information they obtain and rearrange it to acquire and retain new knowledge. On the other hand, constructivism stresses on the notion that students develop further knowledge by building on their previous information and experiences through a sequence of different activities and assessments (Barnard et al. 2009; Shetu et al. 2021). In the DLPCA strategy, in a module-based approach, new information is provided where concepts are connected and constructed from preceding modules. Discussions are around technical topics, as well as practical applications or problems encountered in schools. Assessments are provided to test the students' understanding and problem-solving skills. These strategies provide students with learning because this method also discusses the most applicable learning concepts for the pandemic period. Negovan et al. (2015) found that students, in either face-to-face or distance learning settings, highly consider learning as understanding, which includes boosting one's memorizing, knowledge, and applying what is known. The DLPCA strategy implemented unites these two theories and concepts intending to maximize the role of students in the learning process.

3.1 *Student learning achievement*

School chemistry laboratory management courses were carried out during the pandemic using 100% online learning. All teaching materials and learning instructions are carried out using an LMS developed by the university called SIPEJAR. Students participating in courses have full access as students in SIPEJAR (Surahman et al. 2019). Based on Figure 1, it can be seen that the participants of these studies are spread from all parts of Indonesia. Most of the members in this study took part in learning using notebooks (83%), smartphones (15%), and tablets (3%) (Figure 2). Based on the results of interviews, it is known that lecture participants learning from home (85%), boarding houses (3%), boarding schools (10%), and from other places where there is an internet signal (2%) (Figure 3).

Based on Figure 4, it is known that student learning achievement for one semester following the school chemistry laboratory management course was good. As many as 2% of students got B−, 51% of students got B, 37% of students got B+, and 10% of students got A−. Learning achievement is obtained from discussion scores, structured assignments, midterm exams, and final semester exams with a weighting of 30% for mid-semester exam scores, 30% for end-semester exams, 20% for structured assignments, and 20% for discussion scores.

From the results of questionnaires and interviews, it is known that students who get an A- grade use notebook devices in their learning. During the interview, students admitted that they were more comfortable with a large screen display and enjoyed learning more by repeating the learning videos to better understand the concept of the material. 2% of students who scored C+ used a

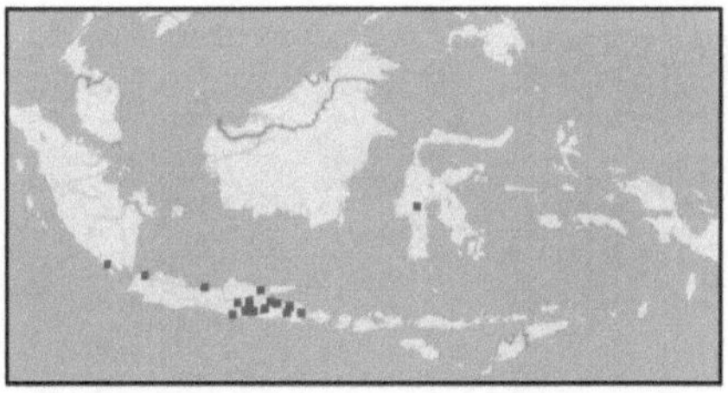

Figure 1. Distribution of participants.

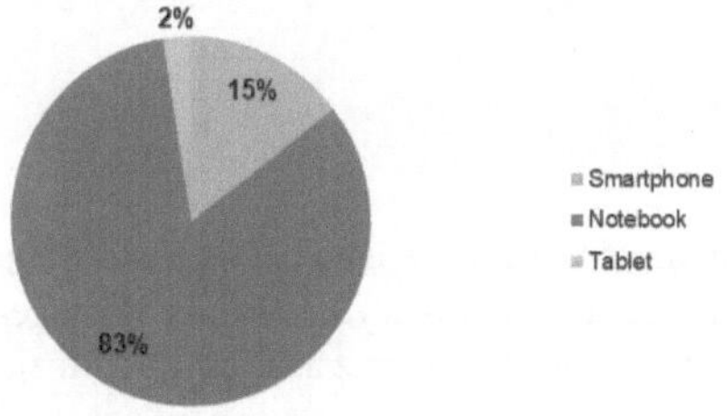

Figure 2. Device of participants.

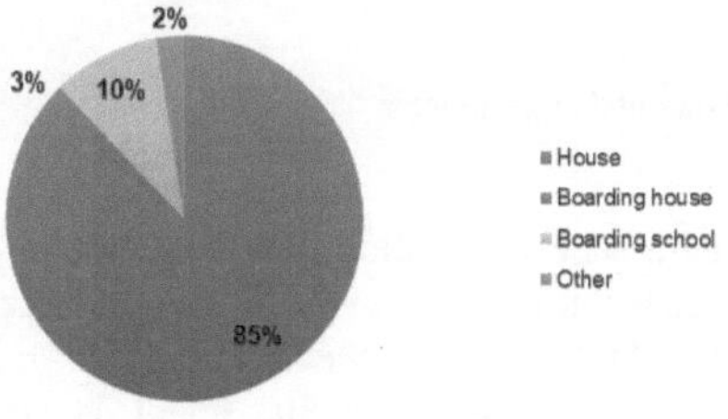

Figure 3. Study site of participants.

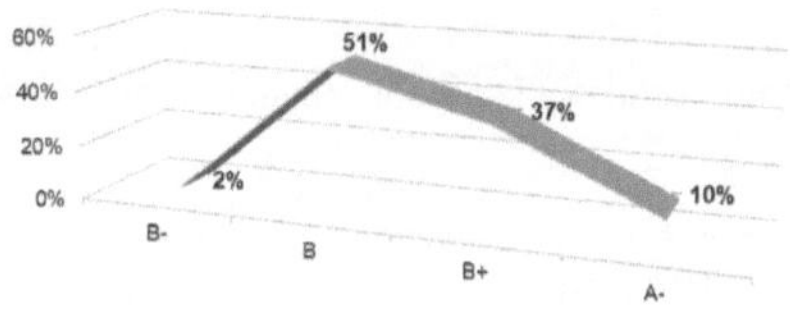

Figure 4. Learning achievement.

smartphone as their device. It is known that they faced difficulty learning by using a smartphone. 15% of students who use smartphones find it difficult to carry out online exams because of the small screen, and at the time of the exam, it is very often that the students answer the questions incorrectly. Around 88% of students who get B and B+ grades use notebooks and smartphones in completing this lecture. From this, we can see that blended learning with the DPLCA strategy positively impacts learning (Hodges et al. 2018; Hoic-Bozic et al. 2016; Lapitan et al. 2021; Seery 2013; Vos et al. 2017).

3.2 *Student learning experience*

At the beginning of the lecture, students are explained the learning objectives and class activities taught with blended learning. Class activities are designed according to the design elements of the chemistry education framework that are in line with the DPLCA strategy. There are some similarities with the lesson plans proposed by Lapitan et al. (2021). The new feature implemented in this unit is an integrated LMS that is recorded on a single server owned by the university. We use a form of cooperative learning to organize the class in active student mode.

From Figure 5, we can see that students carried out the learning activities in 16 meetings with 8 asynchronous meetings and 8 synchronous meetings with an asynchronous time of 100 minutes each. At the discovery stage, students choose modules and videos that have been prepared in SIPEJAR. Students were asked to form group and discuss a video about school chemistry laboratories in Indonesia at the learning stage. From the video, they were asked to identify the advantages and disadvantages of the school chemistry laboratory. Groups that have obtained the advantages and disadvantages of the school chemistry laboratory are asked to determine the parts that should be following applicable regulations and solutions for the school. At the practice stage, the results of group discussions are responded to by other groups in online discussion forums (chat). Each group

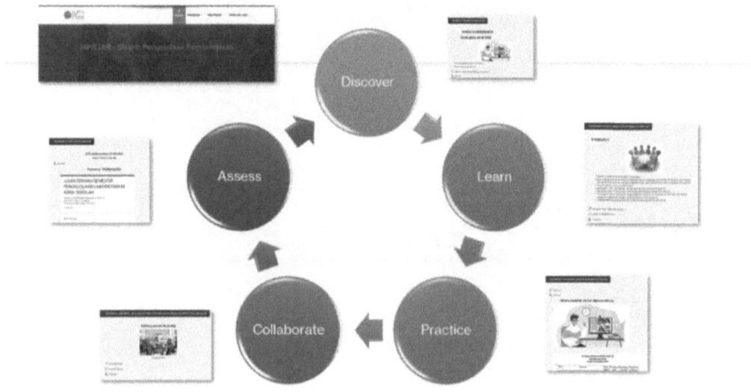

Figure 5. Students' experiences.

will argue with each other about the concepts they understand. At the collaboration stage, each group concludes with what they have understood about the school chemistry laboratory management concepts. At this stage, the lecturer provides reinforcement and synchronization of the concept of school chemistry laboratory management. The last stage is assessing. At this stage, students will be given objective questions and essays to measure their understanding of the concepts of school chemistry management.

Table 1. DPLCA strategy in school chemistry laboratory management courses.

Stage	Description	Activities
Discover	Select our learning online material for each module in the SIPEJAR	• Lecture handouts, course curriculum, lecture videos, and web links to other online resources are provided in SIPEJAR
Learn	The notes and video link for the assigned lecture	• Lecture videos are uploaded in open and free of charge video streaming platform. • Announcements are put up so that the students aware to the time frame
Practice	Apply what you learned using our self-assessment questions	• Detailed explanations problems of chemistry laboratory management • Self-assessment questions are accessible for each topic
Collaborate	Got questions? Get connected and participate in our live online class sessions	• Use of a free of charge and steady video conference platform for online class. • Synchronous meetings are recorded and uploaded in the SIPEJAR. • Conduct of teaching mode of lecture delivery
Assess	Assess your understanding of the topic through exams	• Online assessments are deployed with multiple sets and a specific time to solve and submit the answers.

From the results of questionnaires and interviews, it is known that students strongly agree that a learning video is prepared at the time of learning. With video learning, 85% of students can learn quickly because, during the pandemic, they cannot go to the chemistry laboratory, so they are greatly helped by the visuals of the school chemistry laboratory. Similar research also states that students will more easily understand concepts with the help of visual media (Khery et al. 2021; Prapti Utami & Rohaeti 2019; Sari et al. 2019; Setiawan et al. 2020).

The diagram showed that 72% of students stated that the video quality was quite good, 26% of students noted that the video quality was good, and 1% of students said that the video quality was excellent. If we look at the data distribution of students' responses, it is known that 2% of students stated that they lacked understanding of the concept, 58% of students knew the concept enough, 37% of students understood the concept, and 3% of the students learned the concept very well. From the diagram, it is also known that the ability of students to study independently is as much as

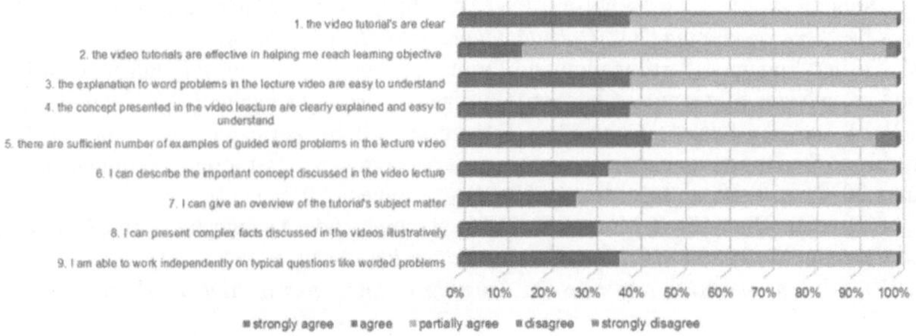

Figure 6. Distribution of students' response.

68% of students are quite capable of independent study, 30% of students can learn independently, and 2% of students are very capable of independent study. Similar research also states that the quality of instructional videos affects students' understanding of learning a concept (Aksela 2005; Prilop et al. 2021).

4 CONCLUSION

This paper presents experiences in implementing the DLPCA strategy, the transition from face-to-face teaching in the classroom to online teaching during the pandemic. The research results show that 90% of students get good grades and 10% of students get very good grades. In addition, as many as 34% of students can study independently. As we can see, this study indicates that the DLPCA strategy can be a constructive inducement to make students learn the management of school chemistry laboratories. This study is one of the first to use LMS (SIPEJAR) in chemistry teaching education in Indonesia. Moreover, educators can use the DLPCA strategy in distance learning, allowing lecture participants to come from various regions or even different countries.

REFERENCES

Aksela, M. (2005). Supporting Meaningful Chemistry Learning and Higher-order Thinking through Computer-Assisted Inquiry: A Design Research Approach. In *University of Helsinki*.

Alshammari, S. H., Ali, M. B., & Rosli, M. S. (2016). The influences of technical support, self efficacy and instructional design on the usage and acceptance of LMS: A comprehensive review. *Turkish Online Journal of Educational Technology*, *15*(2), 116–125.

Ashar, M., Rosyid, H. A., Taufani, A. R., & Nasir, K. R. (2020). Ubiquitous Learning Environment for Smart Improving Disruption Activities in Classroom on Media. *International Journal of Interactive Mobile Technologies (IJIM)*, *14*(13), 200. https://doi.org/10.3991/ijim.v14i13.15609

Backer, J. A., Klinkenberg, D., & Wallinga, J. (2020). Incubation period of 2019 novel coronavirus (2019-nCoV) infections among travellers from Wuhan, China, 20 28 January 2020. *Eurosurveillance*, *25*(5), 1–6. https://doi.org/10.2807/1560-7917.ES.2020.25.5.2000062

Barnard, L., Lan, W. Y., To, Y. M., Paton, V. O., & Lai, S. L. (2009). Measuring self-regulation in online and blended learning environments. *Internet and Higher Education*, *12*(1), 1–6.

Boyle, T., Bradley, C., Chalk, P., Jones, R., & Pickard, P. (2003). Using Blended Learning to Improve Student Success Rates in Learning to Program. *Journal of Educational Media*, *28*(2–3), 165–178.

Chertoff, J. D., Zarzour, J. G., Morgan, D. E., Lewis, P. J., Canon, C. L., & Harvey, J. A. (2020). The Early Influence and Effects of the Coronavirus Disease 2019 (COVID-19) Pandemic on Resident Education and Adaptations. *Journal of the American College of Radiology*, *17*(10), 1322–1328.

Dai, H., & Zhao, B. (2020). Association of the infection probability of CORONAVIRUSwith ventilation rates in confined spaces. *Building Simulation*, *13*(6), 1321–1327. https://doi.org/10.1007/s12273-020-0703-5

Darabi, A., & Jin, L. (2013). Improving the quality of online discussion: the effects of strategies designed based on cognitive load theory principles. *Distance Education*, *34*(1), 21–36.

Gunawan, G., Suranti, N. M. Y., & Fathoroni, F. (2020). *Variations of Models and Learning Platforms for Prospective Teachers During the COVID-19 Pandemic Period. 1*(2), 75–94.

Henry, R. M. (2017). Engaging Participation and Promoting Active Learning through Student Usage of the Internet to Create Notes for General Chemistry in Class. *Journal of Chemical Education, 94*(6).

Hodges, G. W., Wang, L., Lee, J., Cohen, A., & Jang, Y. (2018). An exploratory study of blending the virtual world and the laboratory experience in secondary chemistry classrooms. *Computers and Education, 122*(January 2017), 179–193. https://doi.org/10.1016/j.compedu.2018.03.003

Hoic-Bozic, N., Holenko Dlab, M., & Mornar, V. (2016). Recommender System and Web 2.0 Tools to Enhance a Blended Learning Model. *IEEE Transactions on Education, 59*(1), 39–44.

Inoue, Y. (2010). *Cases on online and blended learning technologies in higher education: concepts and practices*.

Kampf, G., Todt, D., Pfaender, S., & Steinmann, E. (2020). Persistence of coronaviruses on inanimate surfaces and their inactivation with biocidal agents. *Journal of Hospital Infection, 104*(3), 246–251.

Khery, Y., Indah, D. R., Setiawan, N. C. E., Nida, S., Aini, M., Djarwo, C. F., & Nufida, B. A. (2021). Potency of science learning oriented to local wisdom and tourism in enhancing students learning performance. *AIP Conference Proceedings, 2330*(March). https://doi.org/10.1063/5.0043125

Khery, Y., Masjudin, Muzaki, A., Nufida, B. A., Lesnawati, Y., Rahayu, S., & Setiawan, N. C. E. (2020). Mobile-nature of science model of learning for supporting student performance on general chemistry classroom. *International Journal of Interactive Mobile Technologies, 14*(12), 122–137.

Lapitan, L. D., Tiangco, C. E., Sumalinog, D. A. G., Sabarillo, N. S., & Diaz, J. M. (2021). An effective blended online teaching and learning strategy during the COVID-19 pandemic. *Education for Chemical Engineers, 35*(May 2020), 116–131. https://doi.org/10.1016/j.ece.2021.01.012

Marinoni, G., Land, H. Van, & Jensen, T. (2020). *THE IMPACT OF COVID-19 ON HIGHER EDUCATION AROUND THE WORLD IAU Global Survey Report.*

Meltem, E. (2015). The effectiveness of blended learning environments. *Contemporary Issues In Education Research – 4th Quarter, 8*(4), 251–256.

Ministry of Research Technology and Higher Education of the Republic of Indonesia. (2015). Berita Negara. *Ministrial Regulation No. 44/2015 on the National Standards of Higher Education, Nomor 44*(879), 2004–2006.

Negovan, V., Sterian, M., & Colesniuc, G.-M. (2015). Conceptions of Learning and Intrinsic Motivation in Different Learning Environments. *Procedia – Social and Behavioral Sciences, 187*(2004).

Nida, S., Marsuki, M. F., & Eilks, I. (2021). Palm-Oil-Based Biodiesel in Indonesia: A Case Study on a Socioscientific Issue That Engages Students to Learn Chemistry and Its Impact on Society. *Journal of Chemical Education*. https://doi.org/10.1021/acs.jchemed.1c00244

Olakanmi, E. E. (2017). The Effects of a Flipped Classroom Model of Instruction on Students' Performance and Attitudes Towards Chemistry. *Journal of Science Education and Technology, 26*(1), 127–137.

Pienta, N. J. (2016). A "flipped Classroom" Reality Check. *Journal of Chemical Education, 93*(1), 1–2.

Prapti Utami, R., & Rohaeti, E. (2019). Students' Concept Understanding in Chemistry Learning Using Macromedia Flash Based Inquiry Learning. *Online Submission, 10*(3), 1–12.

Prilop, C. N., Weber, K. E., & Kleinknecht, M. (2021). The role of expert feedback in the development of pre-service teachers' professional vision of classroom management in an online blended learning environment. *Teaching and Teacher Education, 99*, 103276.

Rosli, M. S., Saleh, N. S., Aris, B., Ahmad, M. H., Sejzi, A. A., & Shamsudin, N. A. (2015). E-Learning and Social Media Motivation Factor Model. *International Education Studies, 9*(1), 20.

Sari, A. C., Fadillah, A. M., Jonathan, J., & Prabowo, M. R. D. (2019). Interactive gamification learning media application for blind children using android smartphone in Indonesia. *Procedia Computer Science, 157*, 589–595. https://doi.org/10.1016/j.procs.2019.09.018

Seery, M. K. (2013). Harnessing Technology in Chemistry Education. *New Directions, 9*(1), 77–86.

Setiawan, N. C. E., Dasna, I. W., & Muchson, M. (2020). Pengembangan Digital Flipbook untuk Menfasilitasi Kebutuhan Belajar Multiple Representation pada Materi Sel Volta. *Hydrogen: Jurnal Kependidikan Kimia, 8*(2), 107. https://doi.org/10.33394/hjkk.v8i2.3194

Shetu, S. F., Rahman, M., Ahmed, A., Mahin, M. F., Akib, A. U., & Saifuzzaman, M. (2021). Current Research in Behavioral Sciences Impactful e-learning framework?: A new hybrid form of education. *Current Research in Behavioral Sciences, 2*(April), 100038.

Surahman, E., Kuswandi, D., & Wedi, A. (2019). Students' Perception of Project-Based Learning Model in Blended Learning Mode Using Sipejar. *…Conference on Education …* , *372*(ICoET), 183–188.

Vos, J. C., Dragovic, S., Jochimsen, M., Dirach, J., Foth, H., Wiese, J., & Bjerrum, O. J. (2017). From Face-to-Face training to blended learning in the postgraduate program. *European Journal of Pharmaceutical Sciences, 96*, III–VIII. https://doi.org/10.1016/s0928-0987(16)30509-7

Improving Assessment and Evaluation Strategies on Online Learning – Wonorahardjo, Karmina & Habiddin (eds)
© 2022 copyright the Authors, ISBN: 9781032198903

Conceptual development of helper's assessment skills training design on inmate's resilience in L-SIMA

N. Hidayah*, M. Ramli, L. Fauzan, M. Amin & H. Hanafi
Universitas Negeri Malang, Indonesia

ABSTRACT: An inmates life in correctional institution tends to receive negative stereotypes from society in general. This condition became a burden for inmates entering their new life situation. They (inmates) experience stress, depression, and deviant behavior and violations. The ability to surviving their problems and difficulties is required. The resilience of the inmates makes them survive to avoid feeling helpless and hopeless. Based on the explanation of these conditions, the helpers at the Penitentiary Malang (L'SIMA) need to have skills in measuring the resilience condition of the inmates. This study aims to conceptually develop a resilience assessment skills training design for helpers at L'SIMA. This research was conducted using 4D research and development methods with Define Focus, Design and Development Focus, and Disseminate Focus. The main analysis in this study uses form content analysis. The development results are manifested in the design of training procedures and forms of psychological assessment in measuring the resilience of inmates. The design has been developed based on the experiential learning model to provide real and hands-on experience in developing the helpers' skills in measuring the resilience of inmates. Practical testing recommendations are given to find evidence of product effectiveness.

Keywords: Assessment skills, resilience, inmates, helpers

1 INTRODUCTION

Violations of the law occur in many parts of the world in various forms and motives for violations. When an individual commits a violation of the law in the form of a crime, he will get a verdict in the judicial process and serve a sentence in a correctional institution. The purpose of punishment is not only to provide a deterrent effect to perpetrators of criminal acts but rather for self-development to function optimally in society without repeating the same mistakes (Christian et al. 2015).

The coaching process in correctional institutions tends to still get negative stereotypes from society in general (Hamzah & Kumalasari 2018). This situation is not easily accepted by someone with the status of a prisoner, even for those who are still in detention. Correctional institutions are filled with criminals and often do not show conditions that are not conducive to development. This condition is often a burden for inmates who enter their new life situation. It is not uncommon for detainees to experience stress, depression, and exacerbate deviant behavior and previous violations.

The position of detainees as new citizens makes detainees often get bullied by older inmates during their coaching period, either directly or indirectly. This condition certainly does not support coaching the inmates, so extra effort is needed from the inmates to survive in undergoing the coaching period in the correctional institution. Inmates need to endure bullying, persist in improving themselves, and persist in not following other inmates' negative behavior (Ghandehary et al. 2019).

Achieving the goal of returning to normal functioning as community members has many challenges that inmates will face when entering the coaching period. Under these conditions, the

*Corresponding Author

DOI 10.1201/9781003261346-12

resilience of adversity is developed and enhanced by a prisoner. The resilience of these misfortunes is an important condition to be developed from an early age while still serving a period of detention.

The ability to survive in the face of problems and difficulties is necessary for inmates who are just entering the period of formation. The resilience of inmates will make them survive and avoid feelings of helplessness and despair (Markson et al. 2015). The condition of feeling helpless and hopeless is something that needs to be avoided so that individuals will no longer fall into and repeat the same mistakes.

Resilience in this study refers to the strength that individuals show to the system in overcoming adversity. Hence, resilience is the successful adaptation or absence of a pathological outcome after exposure to stressful or potentially traumatic life events or life circumstances. Resilience "combines the interaction of two conditions: risk factors—stressful life events or adverse environmental conditions that increase individual vulnerability—the presence of personal, family and community protective factors that protect, moderate and protect against vulnerability" (Suranata et al. 2017a).

Resilience is the strength to withstand the factors that hinder or interfere with avoiding the risk of negative outcomes. Resilience can guide self-development designed to enhance factors that help detainees cope with the consequences of dangerous risks from their situations and living conditions. In the situation of detainees, resilience is strength to avoid the risk of powerlessness over the negative response of society toward oneself and the determination to develop oneself in preventing the repetition of the same mistakes.

Currently, the resilience of inmates with the label "prospective convicts" is in a vulnerable position. Resilience as a successful adaptation has a big role in building psychological well-being (Nur Hidayah et al. 2020) in inmates. Low and weak resilience tends to make individuals unable to withstand stress, pressure, and face problems (Suranata et al., 2017). On the other hand, high and strong resilience will provide calm, adaptive thinking to lead to the right decision-making in maintaining and improving themselves in the presence of stress, pressure, and problems in their lives (Miranda & Cruz 2020; Santos & Soares 2018).

The magnitude of the conditions of difficulties and problems during the coaching period that will be undertaken is the starting point for the need for resilience in the inmates. These difficulties and problems are also experienced in most correctional institutions, including the Class I Correctional Institution of Malang (L'SIMA). However, currently, there are several obstacles for L'SIMA in identifying the resilience conditions of the inmates. The limitation of instruments is one reason for the delay in identifying the conditions of the detainees.

Based on the explanation of these conditions, the assistants of L'SIMA inmates need to have skills in measuring the resilience conditions of inmates. This is based on the role of facilitators as the party responsible for assisting and monitoring the conditions and adjustment of detainees. The measurement skills of the companions will help understand the conditions of the inmates so that they can provide appropriate assistance in building their resilience during the coaching period at L'SIMA.

Identification of resilience can help facilitators plan the form of assistance provided to inmates and correctional inmates (Hidayah et al. 2021; Segovia et al. 2012). Efforts to identify resilience are expected to maximize the function of resilience as (a) the ability to make decisions to survive, not to give up and not be dragged into a circle of helplessness and despair; (b) the ability to self-regulate on various feelings, especially negative feelings arising from traumatic experiences and problem situations; and (c) the ability to manage thoughts and views or the ability to see better future opportunities.

The process of identifying resilience conditions can be done through a non-test psychological assessment strategy. Non-test psychological assessment in measuring resilience determines the inmates' ability to survive difficult situations and problems they face. Psychological assessments are generally used to identify behaviors, feelings, or actions that cannot be expressed in a single variable or item (Nur Hidayah et al. 2020; Ramli et al. 2020). Using multiple items to measure underlying latent constructs can also explain and isolate certain items' measurement errors (DeVellis 2016). This development process underlies the achievement of more accurate measurement results.

Resilience assessment skills training for inmates L'SIMA Malang assistants is the solution offered to the problems faced by partners. Resilience assessment skills will be a form of accurately identifying the detainees' self-conditions. Thus, appropriate assistance can be provided as a form of follow-up on data on the resilience conditions of the inmates at L'SIMA.

2 METHODS

The research design developed in this study used research and concept development methods that refer to the exposure of 4D research and development procedures by Jerry Wilis (1995; 2000). Research and concept development is descriptive based on a deductive process to find constructions and research foundations in creating an inductive theory of practical problem-solving. Concept research and development is also a basis of urgency for practical research in the form of experiments.

4D research and development consists of the Define Focus, Design and Development Focus, and Disseminate Focus stages. These stages play a role in conceptual and theoretical, reflective construction, and publication. In particular, in the defined focus stages, research activities focus on descriptive analysis in defining the conceptual and theoretical foundations for developing products. References and theoretical literature become research materials to be analyzed through content analysis. The form of content analysis is used to describe the construct and indicators of the success of the product design to be developed.

In the next stage, design and development focus, the development process is based on theoretical construction based on reflection and description in the defined focus stage. This stage will produce theoretical frameworks for designing inmates' resilience assessment skills training for helpers at L'SIMA. The analysis of the content of the procedure is still the main analysis in the reflection of the product development process on the theoretical constructs of the resilience of the inmates.

The third stage of the development process, the dissemination focus stage, becomes a form of publication of product development results. The product development produced in the design phase and the focus of development is a product in the form of a prototype that has met the conceptual and theoretical validity. Dissemination of the results of prototype development is carried out through scientific forums, reports, and research articles.

The participants involved in this development are the Expert Team and the Practitioner Team. The expert participants teams are guidance and counseling experts who play a role in the theoretical reflection of the assessment and the theoretical reflection of resilience. The team of practitioner participants is representatives of L-SIMA helpers who play a role in practical reflection on the resilience conditions of inmates at L-SIMA. Developing and collecting data is carried out through focus group discussion activities in the design and development stage. The instruments used were Focus Group Discussion Guidelines and Focus Group Discussion reflection sheets. The analysis of the development results was carried out reflectively in the reflective conclusion session of the Focus Group Discussion and the descriptive-interpretative analysis of the reflection sheet on the results of the Focus Group Discussion.

3 RESULTS

3.1 *The training procedures*

Copy the template file B1ProcA4.dot to the template directory. This directory can be found by resilience assessment skills training for inmates L'SIMA Malang assistants, which begins with presenting the concepts of designing psychological assessment instruments and the theoretical concepts of resilience and its aspects. This activity is carried out to provide a basis for understanding the concept with the expository method followed by reflective-problematic discussions. The reflective-discussion process on the condition of the inmates is an effort to find alternatives for selecting the most effective forms of assessment to measure the resilience condition of the inmates.

The reflective-problematic discussion that also uses the forms of real experiences of the helpers becomes the initial stage in the experiential learning process, namely concrete experience (Kolb 2015). The identification of experiences in the field conditions of the helpers serves to provide experiences of success and failure in measuring the resilience conditions of the PAPs.

The results of the reflective-problematic discussion were then brought to the stage of reflective observation and abstract construction. The results of the problematic reflective discussion became the main material for internalizing the experience of the theoretical constructs (Nur Hidayah 2012a) of resilience in the inmates. The process is continued in the construction of psychological assessment instruments to measure resilience based on resilience theory. The helpers carried out the main role of constructing the instrument according to the experience data and empirical conditions of the inmates who had high and low resilience.

The next stage of activity focuses on the form of exercise. The first training activity is the design of a psychological resilience assessment instrument. In this activity, the facilitator divides the participants into several groups. Each group is responsible for developing several item constructs from aspects of prisoner resilience, namely (1) Confidence; (2) Emotional Insight; (3) Negative Cognition; (4) Social skills; (5) Connectedness; (6) Supportive Environment; and (7) Availability (Suranata, Atmoko, & Hidayah, 2017b; Suranata, Atmoko, Hidayah, et al., 2017). Design activities are carried out through discussion and group work supervised by the facilitator. The activity continued with the presentation of the discussion results to be reviewed by all participants and facilitators.

The next stage of activity is a simulation of a trial of another report psychological assessment instrument developed. This activity aims to provide a direct experience of the participants as respondents and users of psychological assessment applications. This activity provides an active experimentation stage in experiential learning (Kolb 2015). The instrument testing process will provide hands-on experience and become a form of skill training for helpers in assessing the resilience of the inmates.

The training activity ended with a reflective discussion and evaluation of the results of the activities on the psychological assessment instrument for the resilience of the inmates developed and the results of its testing. The changes experienced by each individual vary according to the experience and conceptualization experienced, even though the training followed is the same (Kolb 2015). The success of the training implementation is influenced by several factors, both internal and external factors. Internal factors come from the participants (companions) and facilitators. This third stage process facilitates these differences through reflection, evaluation, and follow-up on the results of the exercise in designing resilience psychological instruments that have been carried out. The process of reflection, evaluation, and follow-up will also explain the obstacles experienced by the facilitators in implementing the results of this training (Laitila & Oranen 2013). In turn, both the facilitator and the other participants can find solutions to these obstacles.

3.2 *The forms of resilience assessment skills for LSIMA helpers*

The observation technique is a technique of recording individual behavior data through observation by other people either directly and/or indirectly in an activity to obtain an overview of observable behavior. The benefits of observation are that spontaneous behavior data is obtained; naturally, the intensity of behavior is known in detail, and the causes of behavior are known (Ameyaw et al. 2018). Observation techniques need to be equipped with observation instruments, such as Checklist, Rating Scale, Anecdotal Records, Tally, and mechanical devices (Nur Hidayah 2012b).

3.3 *Checklist*

The checklist is an observation record tool containing a list of statements about aspects of a situation, behavior, and activity (individual/group). The checklist administration stage, namely the preparation stage, the implementation stage, and the result analysis stage. The preparation stage includes the following steps: (a) topic determination, (b) variable determination, (c) indicator determination, (d) predictor determination, and (e) statement/item preparation. The implementation phase includes the following steps: (a) preparation of guidelines/DC format, (b) determining the position of observation, namely the observer takes a natural position, (b) knowing the intensity of behavior in detail, and (c) knowing the causes of behavior.

Calculating the percentage (%) with the formula p = N X 100%, then p = 60/90 X 100%, then the result is 66.67%. The next step is to convert the percentage results with the previously created conversion table (cf. Conversion table) so that the results of data interpretation can be concluded.

3.4 *Rating scale*

Rating Scale is an observation record tool containing a list of observable behavior symptoms recorded/checked on a scale. In the observation process with this Rating Scale, the observer records the appearance of behavior based on the scale category. There are three types of scales or degrees of assessment, namely quantitative scale (number scale), qualitative scale (descriptive/word scale), and graphic scale (a combination of number and word scale). Recording the observed behavior symptoms with a Rating Scale, the most important thing is each scale's meaning and description.

Basically, this Rating Scale is useful for self-understanding of the inmates through observation techniques that are more typical of the degree of assessment. The benefits are (a) recording the occurrence of many behaviors systematically, (b) recording the occurrence of many behaviors in a short time, (c) recording the occurrence of many behaviors in the degree of assessment, (d) recording the emergence of behaviors inside and/or outside of school, and (e) recording the occurrence of individual and group behavior at the same time.

3.5 *Tally*

A tally is a tool for recording observations of the appearance of behavior with a vertical line. Tally is useful for recording the intensity of behavior over time and useful for understanding individuals for the benefit of guidance and counseling services. The administration of observations with Tally was carried out in three stages, namely the preparation, implementation, and analysis stages.

3.6 *Mechanical devices*

Mechanical devices are electronic and optical devices used to record data during the observation process. These mechanical tools are usually used as a tool/support data collection with other techniques, such as interviews. Mechanical tools are useful to facilitate or assist in conducting interviews (interviews). Thus, the results of data recording with mechanical devices can complement the data obtained from interviews.

Psychological assessment is divided into two categories based on the data source, namely self-report and other-report (Nur Hidayah 2012b). Other reports are psychological assessments that use other people as data collectors. In contrast, the other type is self-report, a psychological assessment with a source of data coming from the individual concerned. The types of self-report instruments include questionnaires, autobiographies, scales, and interviews. The condition of limited access and the position of detainees in correctional life measures other reports the possible measurement modality to be carried out.

The development of a psychological assessment needs to fulfill several functions (Angelidou et al. 2019), including (1) it is used as an instrument to measure the characteristics of variables and the presentation of variable values ??with scores, (2) is used as an operational definition of abstract concepts, and (3) is used as an instrument for measuring complex or sensitive problems. The accuracy of the instrument development is the basis for the successful collection of data on the resilience of inmates in an accurate manner.

4 CONCLUSION

The measurement skills of the companions will help understand the conditions of the inmates so that they can provide appropriate assistance in building their resilience during the coaching period at L'SIMA. Resilience identification efforts are expected to maximize the resilience function. The process of identifying the condition of resilience can be done through a psychological assessment strategy. The condition of limited access and the position of detainees in correctional life measures other reports the possible measurement modality to be carried out. Resilience assessment skills will be a form of accurately identifying the detainees' self-conditions. Thus, appropriate assistance can be provided as a form of follow-up on data on the resilience conditions of the inmates at L'SIMA. The design of resilience assessment skills training for L'SIMA inmates assistants based on experiential learning is an alternative that can be done. The presence of hands-on experience will help helpers deal with real situations in constructing and implementing the instrument. Product development results are still limited in conceptual validity, and further research is needed for product testing to provide evidence of theoretical validity and practical validity of product effectiveness.

ACKNOWLEDGEMENT

We thank you for the funding from the State University of Malang (represented by LP2M UM). Thank also goes to the Rector of the State University of Malang, who has allowed publication by holding International Conference on Learning Innovation (ICLI) 2021.

REFERENCES

Ameyaw, K. B., Ofori, K. N., Adjepong, B., & Assem, H. D. (2018). Using Observational Techniques as Assessment Instruments to Improve Learners' Performance at Nkawkaw, Ghana. *IOSR Journal of Research & Method in Education*, *8*(6), 87–96.

Angelidou, G., Aguaded-Ramírez, E. M., & Rodríguez-Sabiote, C. (2019). Design and Validation of a Scale Measuring Attitudes toward Refugee Children. *Sustainability*, *11*(10), 2797.

Christian, J., Martinez, D. J., & Martinez, D. (2015). Beyond the Shadows of the Prison: Agency and Resilience Among Prisoners' Family Members. In Joyce A. Arditti & Tessa le Roux (Eds.), *And Justice for All: Families & the Criminal Justice System*. Michigan Publishing, University of Michigan Library.

DeVellis, R. F. (2016). *Scale Development | SAGE Publications Inc* (4th ed., Vol. 26). University of North Carolina.

Ghandehary, M. R. E., Mottaghy Shahri, M. R., & Mohamadi, F. (2019). Mental health status and resilience of female prisoners with an interventional role of physical activity. *Journal of Research & Health*, 488–495.

Hamzah, I., & Kumalasari, F. H. (2018). Self-acceptance and significant others as a factor of the resilience of female prisoners with life sentences. *Journal of Correctional Issues*, *1*(2), 90–99.

Hidayah, N., Hanurawan, F., Ramli, M., Amin, M., & Hanafi, H. (2021). Acceptance and forgiveness contribution to resilience of student's affected by Covid-19. *Turkish Journal of Physiotherapy and Rehabilitation*, *32*, 1130–1137.

Hidayah, Nur. (2012a). Supervisi Klinis Berbasis Experiential Learning pada Praktik Konseling. *Psikovidya*, *16*(2), 15–22.

Hidayah, Nur. (2012b). *Teknik Pemahaman Individu*. Universitas Negeri Malang.

Hidayah, Nur, Hanafi, H., Ramli, M., & Rofiqoh. (2020). Student's Meaning of Life Based on Resilience in New Normal Era. *1 St International Conference on Information Technology and Education (ICITE 2020)*, 239–244.

Hidayah, Nur, Ramli, M., Mappiare-AT, A., Hanafi, H., Yuliana, A. T., Kurniawan, N. A., & Eva., N. (2020). Developing Critical Thinking Skills Test In Indonesia. *Palarch's Journal Of Archaeology Of Egypt/Egyptology*, *17*(3), 815–826.

Kolb, D. A. (2015). *Experiential Learning: Experience as a Source of Learning and Development*. Prentice Hall. http://www.sciepub.com/reference/245957

Laitila, A., & Oranen, M. (2013). Focused Dialogues in Training Contexts?: A Model for Enhancing Reflection in Therapist ' s Professional. *Contemporary Family Therapy: An International Journal*, *35*(3), 599–612.

Markson, L., Lösel, F., Souza, K., & Lanskey, C. (2015). Male prisoners' family relationships and resilience in resettlement. *Criminology & Criminal Justice*, *15*(4), 423–441.
Miranda, J. O., & Cruz, R. N. C. (2020). Resilience mediates the relationship between optimism and well-being among Filipino university students. *Current Psychology*, 1–10.
Ramli, M., Hidayah, N., Eva, N., Nor, D. M. B. M., Saputra, N. M. A., & Hanafi, H. (2020). The Counselors' Need for the Development of A Solution-Focused Cybercounseling Model for Junior High School Students. *2020 6th International Conference on Education and Technology (ICET)*, 209–213.
Santos, Z. de A., & Soares, A. B. (2018). Social skills, coping, resilience and problem-solving in psychology university students. *Liberabit: Revista Peruana de Psicología*, *24*(2), 265–276.
Segovia, F., Moore, J. L., Linnville, S. E., Hoyt, R. E., & Hain, R. E. (2012). Optimism predicts resilience in repatriated prisoners of war: A 37-year longitudinal study. *Journal of Traumatic Stress*, *25*(3), 330–336.
Suranata, K., Atmoko, A., & Hidayah, N. (2017a). Validation of Psychological Resilience Youth Developement Module and Implication for Guidance and Counseling. *GUIDENA: Jurnal Ilmu Pendidikan, Psikologi, Bimbingan Dan Konseling*, *7*(1), 57.
Suranata, K., Atmoko, A., & Hidayah, N. (2017b). Enhancing Students' Resilience: Comparing The Effect of Cognitive-Behavior And Strengths-Based Counseling. *Proceedings of the 2nd International Conference on Innovative Research Across Disciplines (ICIRAD 2017)*, 102–108.
Suranata, K., Atmoko, A., Hidayah, N., Rangka, I. B., & Ifdil, I. (2017). Risks and resilience of students with hearing impairment in an inclusive school at Bengkala, Bali, Indonesia. *Specialusis Ugdymas*, *2*(37), 165–214.
Willis, J. (1995). A recursive, reflective instructional design model based on constructivist-interpretivist theory. *Educational Technology*, *35*(6), 5–23.
Willis, Jerry, & Wright, K. E. (2000). A General Set of Procedures for Constructivist Instructional Design: The New R2D2 Model. *Educational Technology*, *40*(2), 5–20.

Improving Assessment and Evaluation Strategies on Online Learning – Wonorahardjo, Karmina & Habiddin (eds)
© 2022 copyright the Authors, ISBN: 9781032198903
Open Access: www.taylorfrancis.com, CC BY-NC-ND 4.0 license

"She is a support system": Peer assessment in cooperative learning asynchronous classroom

Sari Karmina*, Ahmad Heki Sujiatmoko, Lestari Setyowati & Bella Indah Lestari
Universitas Negeri Malang, Indonesia

Wen Xiong
University of Auckland, New Zealand

ABSTRACT: Peer assessment is essential in cooperative learning in asynchronous classroom to help teachers evaluate the implementation of cooperative learning. However, little research on peer assessment in cooperative learning asynchronous classroom has been reported. Thus, the study aims to examine the implementation of cooperative learning through peer assessment in cooperative learning asynchronous learning. Thirty-two students participated. The study used a quantitative research approach employing closed and open-ended questionnaires. The closed-ended questionnaire was analysed quantitatively, and the open-ended questionnaire was analysed using thematic analysis. The findings from the closed-ended questionnaire revealed that more than 50% of the students actively contributed to group goals, showed consideration of others, and helped groups identify changes and encouraged group actions. Three themes were generated from the open-ended questionnaire: peer roles, peer dispositions, and peer contribution. The study suggests that peer assessment be an effective tool to assess the implementation of cooperative learning in asynchronous learning.

Keywords: Asynchronous, EFL, peer assessment, student-centred learning

1 INTRODUCTION

Due to the Covid-19 outbreak, a new policy in the education sector has been set and implemented in the teaching and learning system in Indonesia. Teachers and students from any levels of education are regulated to conduct the teaching and learning processes via online since March 2020. Classes have been conducted online: synchronously and asynchronously. Synchronous learning means that when students learn from a distance; they virtually attend or join a class session (Bower et al. 2015; Chen et al. 2005). A teacher and students meet through video conferencing platforms, such as Zoom, Google Meet, etc. There are direct learning and reciprocity in a lesson that has been mutually agreed upon in advance. On the other hand, asynchronous learning usually allows students to study on their own schedule, within a certain time (Johnson 2006). The students can access and complete lectures, readings, homework and other learning materials at any time. A big benefit to asynchronous classes is the flexibility. Asynchronous classes mean that the students do not always need to be online at the same time as their teacher or classmates (Scheiderer 2021).

Previous research reports that online learning has caused positive and negative impacts (Chen et al. 2005; Dhawan 2020; Nieuwoudt 2020; Simamora 2020; Yamagata-Lynch 2014). Researchers report that online learning offers students freedom to manage time, provide more opportunities to collaborate with other students, and to explore deeper on the content (Chen et al. 2005; Dhawan 2020; Nieuwoudt 2020). On the other hand, online learning has been described as bringing negative impact towards the students (Dhawan 2020; Simamora 2020). Simamora (2020) reported that the

*Corresponding Author

 DOI 10.1201/9781003261346-13

students experienced health problems like fatigue and headache, difficulties in financial to buy online quota, and difficulties to engage in a lesson with a particular topic with various learning approaches from teachers (Simamora 2020).

Online learning requires technology readiness (Chen et al. 2005; Dhawan 2020) and careful planning to engage students (Yamagata-Lynch 2014). Regarding this, there are two things that must be considered by the teachers in conducting or holding the online or virtual learning activities. First is concerned with the existence and provision of technology as the facility to support the implementation of the online or virtual classrooms. In this case, the online or virtual learning activities may work well and successfully if they are supported with the use of technology which can provide many facilities. For example, the online or virtual learning activities needs to use some digital applications to be connected with internet connection to provide the access for both the teacher and students to have the virtual meeting. Such available digital applications used can be like google meet room or zoom. Second is concerned with the teaching and learning strategies which can create certain attractive learning situations to support the implementation of the online or virtual classrooms.

One of the instructional strategies that engage the students during online learning is cooperative learning (Ivone et al. 2020; Jacobs & Seow 2015). Yamagata-Lynch (2014) recommended that teachers need to conduct small group meetings to help students feel connected with their friends with the support of the teachers. In addition to that, Ivone, Jacobs, and Renandya (2020) suggested that cooperative learning be used to provide students with cooperative skills and communication technology skills. Jacobs and Seow (2015) proposed eight principles to improve students' interaction in online setting: heterogenous grouping, teaching collaborative skills, maximum peer interactions, equal opportunity to participate, individual accountability, positive interdependence, and cooperative as a value. Jacobs and Seow (2015) suggested that, in addition to the principles teachers need to be familiar with the information and communication technologies and collaborative application such as Google Docs and collaborative mind mapping applications. Similar to Jacobs and Seow (2015), drawn from the theory of social interdependence theory, Johnson and Johnson (2009) proposed five principles of cooperative learning: positive interdependence, individual accountability, promotive interactions, appropriate use of social skills, and group processing. Without the establishment of cooperative learning principles, it is hard to carry out cooperative learning successfully (Johnson & Johnson 2009).

However, research shows that online cooperative learning lacks students' engagement (Henry et al. 2015; Junco et al. 2017). Sukmawati et al. (2020) reported that some students do not engage in the class, and some may not be available to meet at specific times when they have synchronous or asynchronous classes. Hence, it is crucial for the teachers to have alternative assessment which can inform the students' involvement in cooperative learning activities. One of the assessments that may provide the access for knowing how the students get involved in the cooperative learning group is peer assessment. Peer assessment is supposed to be one of the authentic assessments that promotes the way to measure the individual's characteristics in doing the learning activities (O'Malley & Pierce 1996). In cooperative learning, peer assessment may contribute to give the information about the individuals' engagement as their depiction of the seriousness of reaching the cooperative attitudes as opposed to the competitive learning (Lejk & Wyvill 2001; Orsmond et al. 1996). Research suggests that peer assessment in cooperative learning limit free-riders or hitchhikers in a group task (Ayadi & Ayadi 2009). However, little research has been reported about peer assessment in cooperative learning asynchronous classroom particularly in the Indonesian context. Thus, this current study aims to investigate the implementation of cooperative learning in asynchronous classroom through peer assessment.

2 METHODS

The current study aims to examine the students' implementation of cooperative learning in asynchronous learning through peer assessment. The current study employed a mixed-method research approach using closed- and open-ended questionnaire as data collection method to answer the research question (see Creswell, 2012). A research question is addressed: do the students implement cooperative learning in asynchronous classroom?

2.1 *Context and participant*

The current study was conducted in an EFL (English for Foreign Language) synchronous and asynchronous classroom at the Department of English, Universitas Negeri Malang. Thirty-two Semester 4 students joined the study. The students studied in 6 meetings synchronously and 28 meetings asynchronously. English language was used as a means of communication because the students majored in English Language and Literature. At the first meeting, the students were given a synchronous workshop on how to work in CL group. The workshop discussed CL philosophies, cooperative learning principles, and cooperative learning structures. The orientation of cooperative learning was crucial to train the students cooperative learning principles (see Karmina et al. 2021). The students stayed in one group for 4–6 asynchronous meetings then they changed partners after a new topic. The group was formed by a learning management system, so that everyone in the class would have a chance to work with different people. There was, however, a chance for the students to be in the same group two or three times during the course of the study.

2.2 *Data collection and analysis*

The research instrument of the current study was a closed and open-ended questionnaire (see Cresswell 2012). The questionnaire was administered through *Google Form,* a free online survey that enables users to create surveys and receive responses and it provides the analysis of the survey (Mondal et al. 2018). Each student in the current study assessed all of his/her group members. Thus, each student filled out at least two or three forms. Prior to giving the questionnaire, the authors explained the items and how to answer the items to avoid misunderstanding.

Four closed-ended items and one open-ended question were adapted from cooperative learning rubric developed by International Reading Association and National Council of Teachers of English (2003). The rubric represents the five principles of cooperative learning (see Johnson & Johnson 2009). In addition, the rubric was selected because it worked for the context of the study. The students were asked to choose which number best represents their friend when working in the group.

The data of the closed-ended questionnaire were automatically analysed by Google Form in the form of percentage. In addition to that, the researchers conducted manual calculations on Microsoft Excel to confirm the results from the Google Form. The data from an open-ended question were analysed using thematic analysis (Braun & Clarke 2006). Braun and Clarke (2006) proposed six phases of thematic analysis: data familiarisation, initial code generation, theme search, theme review, theme naming, and reporting. In the current study, the data familiarisation was conducted by reading and re-reading the data and noting down initial ideas. After that, the researchers coded the entire data set and collated data which had similar codes. The third phase was searching for themes by collating codes into potential themes. Then, the researchers checked if the themes represented the coded chunks of data. At the fifth stage before reporting, the themes were defined and named. The final stage is reporting the results of the data analysis.

3 RESULTS AND DISCUSSION

The close-ended questionnaires asked the students how they assessed their team based on their contribution to the group goals, consideration of others, contribution of knowledge, and working and sharing with others. The results of the first item, contribution to group goal, is described in Table 1.

Table 1 shows how the team members practised at least 2 cooperative learning principles: positive interdependence and individual accountability (see Johnson & Johnson 2009). Positive interdependence occurs when the students encourage cooperation among students and help each other boost their achievement (Kagan & Kagan 2009). Individual accountability is seen when the group assists its members to be a strong member (Dyson & Casey 2012). The results reveal that the majority of the students committed to the last category in which the students had a good awareness of working

Table 1. Contribution to group goals.

Category	Percentage
1. Works towards group goals when prompted.	7.5%
2. Works towards group goals with occasional prompting.	9.3%
3. Works towards group goals without occasional prompting: accepts and fulfils individual role within the group.	18.7%
4. Consistently and actively works towards group goals; willingly accepts and fulfils individual roles within the group.	64.5%

together by accepting and fulfilling each role that had been decided as the jobs assigned. These findings were supported by the results of the open-ended questionnaire when they reported that the team members had different roles. Some of the students' comments are:

> "She's like the leader of the group. She [is] often being the one who reminds us to work the group assignment."
> "She's also the secretary of the group because she is willing to change the answer whenever there are changeable answers."
> "She is a support system. She laughs at our jokes and encourage[s] us by saying *semangat* [we can do it]."
> "He [is] always be the host who creates the Zoom meeting."
> "She is the mood maker for me."

The division of roles in the group increases the positive interdependence and individual accountability of the members (Gillies 2007; Johnson & Johnson 2009). Each member is responsible for his/her own task to achieve the group's goal.

The results of Table 2 also reveal that most students were engaged to give contributions to reach the group's goal. This finding is in line with Yamagata-Lynch's (2014) that the students would be well engaged if the group work is well-structured in synchronous or asynchronous classroom. The teamwork could not be separated from the acts of being aware of what each group member needed to do. What happened with the teamwork was always related to the way how to reach the same goal and success (see Gillies 2007; Johnson & Johnson 2009). One student who does not have a good participation in doing the jobs assigned will make the team difficult to fulfil what is expected. Hence, the acts of participating in doing the jobs assigned is essential to reach the goal and success expected in the teamwork.

Table 2. Consideration of others.

Category	Percentage
1. Needs occasional reminders to be sensitive to the feelings of others.	9.3%
2. Shows sensitivity to the feelings of others.	15%
3. Shows and expresses sensitivity to the feelings of others; encourages the participation of others.	21.5%
4. Shows sensitivity to the feelings of others; values the knowledge, opinion, and skills of all group members.	54.2%

Table 2 represents the students' implementation of promotive interaction and appropriate use of social skills principles. Most of the students were not only sensitive to the feelings of others, but also, they valued the knowledge, opinions, and skills of others. When doing the cooperative learning, most students in the group tended to keep their own feelings for respecting others who shared opinions and showed their skills. These findings are supported by the students' comments of their team members from the open-ended questionnaire. One student wrote, "I think she is quite active in the group discussion. I often have different answers with her but she [is] always be open-minded to deal with it." Another student commented on her team member, "I think she is a good member as she is very considerate toward others." Open-minded and considerate are some of the dispositions that a group member should have since cooperative learning team constantly discusses the task to achieve the goal. There will be many disagreements during the discussion. Here are some other excerpts how the students promote positive interactions and use appropriate skills.

> "I think that she always respects when other members of our group give an opinion."
> "She is always active and leads the discussion most of the time along with George [pseudo name]."
> "He sometimes has different opinions from us, but his different opinion leads us to the better discussion. It's good."
> "She is fun to discuss with because sometimes we are not in the same page, and she's quite attentive as she often notices the little mistakes in our group."

The students' attempt on promoting positive interactions and practising the use of appropriate skills are in line with previous research findings that cooperative learning creates positive outcomes of social interdependence (Cohen & Lotan 2014; Johnson & Johnson 2009).

Table 3. Contribution of knowledge.

Category	Percentage
1. Contribute information to the group only when prompted.	5.6%
2. Contributes information to the group with occasional prompting and reminding.	10.3%
3. Contributes knowledge, and skills without prompting or reminding.	38.3%
4. Consistently and actively contributes knowledge, opinions, and skills without prompting and reminding.	45.8%

The results of Table 3 shows that most of the students were individual accountable. The students tended to contribute knowledge, opinions, and skills without prompting and reminding. The finding shows that the students share responsibility to complete the group's goal. This finding is supported by the open-ended question data: peer contribution theme.

> "I think she did a great job on most of the discussion. She contributed well to giving her opinion during discussion."
> "Honestly, my teammates are very helpful and all of us contributed to the group discussion actively and passionately."
> "She paid attention to other's answers in the group and seemed to not hesitate to ask or give suggestions."
> "Her understanding of the material is remarkable. She knows very well how to explain the materials to the others."

The shared responsibility that the students created increased the feeling of accountable and motivation to perform well (see Gillies 2007). The current study findings confirm Divaharan and Atputhasamy's (2002) research. Divaharan and Atputhasamy's (2002) students reported that through peer assessment they learned to be responsible in working in cooperative learning group.

Table 4. Working and sharing with others.

Category	Percentage
1. Participate in needed changes when promoted and encouraged; always or often relies on others to do the work.	7.5%
2. Participate in needed changes with occasional prompting; often needs reminding to do the assigned work.	10.3%
3. Willingly participates in needed changes; usually does the assigned work and rarely needs reminding.	29%
4. Helps the group identify necessary changes and encourages group action for change; does assigned work without reminders.	53.3%

Table 4 shows that most of the students practiced positive interdependence by doing the assigned work without being reminded, and they conducted group processing by helping the group identify necessary changes and encourages group action for change, and the students. Group processing is important in cooperative learning group. Each member should be able to ascertain that cooperation among the individuals in the group may work well. This can happen if each member has high social care and attention to each other. Some students commented:

> "He is the one who often makes the group reflection."
> "She is such a supportive friend. She always checks up on others."
> "He is very cooperative and always active on group work. He also tried to check up on the others."

Group processing allows the team to evaluate their performance, and it is an opportunity for the team to improve their social and emotional skills (Dyson et al. 2016).

4 CONCLUSION

Online learning is inevitable in this COVID-19 pandemic. Cooperative learning is an alternative approach to teaching and learning that fosters cooperation and collaboration in asynchronous learning. The absence of teachers' supervision in asynchronous learning makes it difficult for the teachers to evaluate whether students implement cooperative learning. Thus, the current study aims at investigating the students' implementation of cooperative learning principles during asynchronous classroom through peer assessment. The findings reveal that peer assessment is an effective tool to assist teachers to find out if the students implement cooperative learning principles in asynchronous learning. The current study provides a significant contribution to the study of peer assessment in asynchronous cooperative learning classroom by offering an alternative way of monitoring the students' practice of cooperative learning principles. Moreover, the findings of the current study indicate that peer assessment provides an authentic assessment of how the students feel about working in the group and how they feel about their team. This assessment is important for teachers to make changes in their teaching.

ACKNOWLEDGEMENT

The authors would like to acknowledge the Faculty of Letters and the Research and Community Service Unit of Universitas Negeri Malang, Indonesia for funding the current research.

REFERENCES

Aguilera-Hermida, A.P. 2020. College students' use and acceptance of emergency online learning due to COVID-19. *Int. J. Educ. Res. 1*, 100011.

Ayadi, M. F., & Ayadi, O. F. 2009. Implementing Peer Evaluation to Enhance a Cooperative Learning Environment. *Learning in Higher Education*, 15.

Bower, M., Dalgarno, B., Kennedy, G. E., Lee, M. J., & Kenney, J. 2015. Design and implementation factors in blended synchronous learning environments: Outcomes from a cross-case analysis. *Computers & Education, 86*, 1–17.

Braun, V., & Clarke, V. 2006. Using thematic analysis in psychology. *Qualitative Research in Psychology*, 3(2), 77–101. doi:10.1191/1478088706qp063oa

Careaga-Butter, M., Quintana, M. G. B., & Fuentes-Henríquez, C. 2020. Critical and prospective analysis of online education in pandemic and post-pandemic contexts: Digital tools and resources to support teaching in synchronous and asynchronous learning modalities. *Aloma: revista de psicologia, ciències de l'educació i de l'esport Blanquerna, 38*(2), 23–32.

Chen, N. S., Ko, H. C., Kinshuk*, & Lin, T. 2005. A model for synchronous learning using the Internet. *Innovations in Education and Teaching International, 42*(2), 181–194.

Cohen, E. G., & Lotan, R. A. 2014. *Designing groupwork: Strategies for the heterogeneous classroom* (3rd ed.). New York, NY: Teachers College Press.

Creswell, J. W. 2012. *Educational research: Planning, conducting, evaluating quantitative and qualitative research* (4th ed.). Boston, MA: Pearson.

Dhawan, S. 2020. Online learning: A panacea in the time of COVID-19 crisis. *Journal of Educational Technology Systems, 49*(1), 5–22.

Divaharan, S., & Atputhasamy, L. 2002. An attempt to enhance the quality of cooperative learning through peer assessment. *The Journal of Educational Enquiry*, *3*(2).
Dyson, B., & Casey, A. (Eds.). 2012. *Cooperative learning in physical education: A research-based approach*. Milton Park, England: Routledge.
Dyson, B., Colby, R., & Barratt, M. 2016. The co-construction of cooperative learning in physical education with elementary classroom teachers. *Journal of Teaching in Physical Education*, *35*(4), 370–380. doi:10.1123/jtpe.2016-0119
Gillies, R. M. 2007. *Cooperative learning: Integrating theory and practice*. Thousand Oaks, CA: Sage.
Henrie, C. R., Halverson, L. R., & Graham, C. R. 2015. Measuring student engagement in technology-mediated learning: A review. *Computers & Education*, 90, 36–53.
International Reading Association & National Council for Teacher Education. 2003. *Cooperative learning rubric*. https://www.readwritethink.org/sites/default/files/resources/lesson_images/lesson95/coop_rubric.pdf
Ivone, F. M., Jacobs, G. M., & Renandya, W. A. 2020. Far apart, yet close together: Cooperative learning in online education. *Studies in English Language and Education*, *7*(2), 271–289.
Jacobs, G., & Seow, P. 2015. Cooperative learning principles enhance online interaction. *Journal of International and Comparative Education (JICE)*, 28–38.
Johnson, G. M. 2006. Synchronous and asynchronous text-based CMC in educational contexts: A review of recent research. *TechTrends*, *50*(4), 46–53.
Johnson, D. W., & Johnson, R. T. 2009. An educational psychology success story: Social interdependence theory and cooperative learning. *Educational Researcher*, 38(5), 365–379. doi:10.3102/0013189X09339057
Junco, R., Heiberger, G., & Loken, E. 2011. The effect of Twitter on college student engagement and grades. Journal of computer assisted learning, 27(2), 119–132
Kagan, S., & Kagan, M. 2009. *Kagan cooperative learning*. San Clemente, CA: Kagan Publishing.
Karmina, S., Dyson, B., Watson, S. J., Penelope, W., & Philpot, R. 2021. Teacher implementation of cooperative learning in Indonesia: A Multiple Case Study. *Education Sciences*, *11*(5), 218.
Khan, A., Egbue, O., Palkie, B., & Madden, J. 2017. Active learning: Engaging students to maximize learning in an online course. *Electronic Journal of E-Learning*, 15(2), pp107–115.
Kahn, P., Everington, L., Kelm, K., Reid, I., & Watkins, F. 2017. Understanding student engagement in online learning environments: The role of reflexivity. *Educational Technology Research and Development*, 65(1), 203–218.
Lejk M & Wyvill M. 2001. Peer assessment of contributions to a group project: a comparison of holistic and category-based approaches. Assessment and Evaluation in Higher Education, vol 26, no 1, pp 61–72.
Meulenbroeks, R. 2020. Suddenly fully online: A case study of a blended university course moving online during the Covid-19 pandemic. *Heliyon*, *6*, e05728.
Mondal, H., Mondal, S., Ghosal, T., & Mondal, S. 2018. Using Google forms for medical survey: A technical note. *Int J Clin Exp Physiol*, 5(4), 216–218.
Nieuwoudt, J. E. 2020. Investigating synchronous and asynchronous class attendance as predictors of academic success in online education. *Australasian Journal of Educational Technology*, *36*(3), 15–25.
Orsmond P, Merry S & Reiling K. 1996. The importance of marking criteria in the use of peer assessment. Assessment and Evaluation in Higher Education, vol 21, no 3, pp 239–250.
O'Malley, J. Michael & Lorraine. V. Pierce. 1996. *Authentic assessment for English language learners: Practical approaches for teachers*. New York: Addison-Wesley Publishing Company.
Scheiderer, J. (2020). *What's the difference between asynchronous and synchronous learning?*. Ohio State Online. Retrieved from https://online.osu.edu/resources/learn/whats-difference-between-asynchronous-and-synchronous-learning
Schilstra, T., Takács, E., & Abcouwer, T. 2019. Cooperative learning in a higher educational setting: Realizing high performing cooperative learning in higher education. In *Proceedings of the 2019 Association for Information Systems (AIS SIGED) International Conference on Information Systems Education and Research* (Vol. 7).
Simamora, R. M. 2020. The Challenges of online learning during the COVID-19 pandemic: An essay analysis of performing arts education students. *Studies in Learning and Teaching*, *1*(2), 86–103.
Sukmawati, R. A., Pramita, M., Purba, H. S., & Utami, B. 2020. The use of blended cooperative learning model in introduction to digital systems learning. *Indonesian Journal on Learning and Advanced Education (IJOLAE)*, *2*(2), 75–81.
Yamagata-Lynch, L. C. 2014. Blending online asynchronous and synchronous learning. *International Review of Research in Open and Distributed Learning*, *15*(2), 189–212.

Improving Assessment and Evaluation Strategies on Online Learning – Wonorahardjo, Karmina & Habiddin (eds)

The effectiveness of contextual problem–solving-based integrative online learning on fundamental physics for higher education

Y. Affriyenni*, S. Sutopo & C.I. Yogihati
Universitas Negeri Malang, Indonesia

ABSTRACT: The prolonged COVID-19 pandemic forces learning to be conducted online using technology. This study aims to investigate the effectiveness of problem–solving-based learning using Moodle-based Learning Management System (LMS) integrated with synchronous online meeting and virtual laboratory towards students' physics comprehensive understanding and problem-solving skills compared to problem–solving-based learning that only uses Moodle-based LMS as either the synchronous and asynchronous platform. This study was conducted by using a pretest-posttest quasi-experiment design supported by qualitative response involving 66 students selected randomly in Science Education Study Program, Faculty of Mathematics and Natural Sciences, Universitas Negeri Malang. This study shows that students who took a problem–solving-based online course using synchronous online meeting and virtual laboratory integrated LMS, their physics problem-solving skills and comprehensive conceptual understanding are better than those who learn only using non-online meetings synchronous such as only using discussion forum and assignment. Besides, students were more enthusiastic and more active in discussing briefly asynchronously before the online meeting session. Students' fundamental concepts are better constructed when they are involved in the investigation using the virtual laboratory to solve contextual problems. Hence, the use of LMS integrated with online meeting platforms and virtual laboratories is effective to improve physics understanding comprehensively and develop students' contextual problem-solving skills. This study has the potential to be developed further for wider use in learning.

Keywords: Problem-solving, Integrative online learning, Fundamental Physics

1 INTRODUCTION

Education in the 21st century integrates knowledge, skill, and attitude competence along with technical proficiency. Science education is directed to prepare students to be successfully live in the century (Liu 2009). One of the skills necessary is science literacy. Science literacy requirements in the national and international community arise because every individual needs to participate to solve real-life problems through science and technology mastery based on mathematics, physics, chemistry, biology, and environmental understanding (Cardwell 2005). It is a special concern to prepare the pre-service teacher to have the skill in developing students' capability to face the changes in the 21st century (Gultepe & Kilic 2015).

One of the current real-time challenges in conducting the learning process is how to develop students' skills during the COVID-19 pandemic. Currently, Indonesia is in a national emergency with the highest death rate in the world. Changes and policy renewal are necessary including in the education sector (Sukmana et al. 2020). The education sector needs to change its initial learning technique from offline meetings to online meetings following the physical distancing policy (Abidah et al. 2020; Sun et al. 2020). This policy has been conducted for two years where all academic communities keep making adjustments in the process. Online learning is an alternative learning method that

*Corresponding Author

DOI 10.1201/9781003261346-14

utilizes the virtual world in the learning process (Goldschmidt 2020). Various platforms are available to support online learning such as chat applications, LMS, meeting conferences, simulation, and other graphical applications. Physics learning is no exception in using the online platform.

Physics learning faces a huge challenge since its characteristic is deemed as a challenging course even when the learning process was conducted offline. The learning process especially physics should not only be conducted by text but media to facilitate the comprehensive conceptual acquisition. Besides, related to the character as one of the science courses, physics learning should make students capable to apply the concept to problem-solving. Hence, students should be accustomed to using various representations (Treagust et al. 2017). As a pre-service science teacher, students should be accustomed to following physics learning directed to comprehensive conceptual understanding-based problem-solving to be able to conduct such a learning process later in school. Hence, online physics learning by using a learning management system (LMS)'s facilities such as discussion forum and assignment or chat application is not enough but needed to be supported by other means of learning platform to make a meaningful impact on students. One of the means is using practical simulation to temporarily substitute experiment that is supposedly conducted offline. Besides, synchronous conceptual reinforcement using an online meeting conference platform could simplify the learning process by providing direct interaction between the lecturer and the students. No existing platforms could give optimal impact if conducted partially without integration. Thus, this study aimed to identify the effectiveness of integrative online platforms towards problem–solving-based fundamental physics II courses.

2 METHODS

This study was conducted using the quasi-experimental method with two groups consisting of the control and experiment groups. The study was conducted towards the first-year pre-service science teacher in the science education study program. The intended research sample was students who took the Fundamenta Physics II course in the first year's second semester of the 2020/2021 academic year. The control group consisted of 33 students who learned using a problem–solving-based approach using the discussion, assignment, and quiz features on the LMS platform and received several other instructions through the chat application. The experimental group consisted of 33 students who learned using the problem-solving approach using LMS integrated with *Google Meet* conference meeting, *PhET simulation, WhatsApp* chat application, and LMS feature itself including the discussion, assignment, and quiz. To minimize research variability, several initial factors of each group were controlled including initial knowledge, age, and learning achievement of the prerequired course (Fundamental Physics I). These factors were determined by collecting information about students' average age, investigating learning achievement of the prerequired course, examining students' pre-test answers, and determining students' scores.

The quantitative data was obtained based on the test and questionnaire. The test used in this research was related to problem-solving that needs comprehensive conceptual understanding before and after treatment. The treatment was conducted in 16 meetings. Meanwhile, the questionnaire was used to identify students' responses toward the use of the integrative online platform. The questionnaire used the Likert 4 scales. The used questions and questionnaire had been discussed by the teaching team where the used questions had been validated and reliability tested continuously for each development.

The question set consists of 40 HOTS items with the blueprint as shown in Table 1. Moreover, qualitative data was acquired using a questionnaire with the blueprint and response is shown in Table 2. The observation was conducted based on students' activity in the discussion forum and on the synchronous session on the *Google Meet* platform. Besides, students' suggestions were also received at the end of the questionnaire. Meanwhile, the conducted treatment on the control and experimental group is shown in Table 3.

Based on the quantitative data in the test, the mean scores of control and experimental groups were compared using independent samples t-test. Afterward, the *N-Gain* score was also conducted.

The independent samples t-test was used to test if there are any differences between the control and experimental group. Meanwhile, the *N-Gain* score test was conducted to find the effectiveness of used learning media. The categorization of *N-Gain* scores is shown in Table 4. This study used statistical software to analyze the learning achievement of the control and experimental group and to test the effectiveness of the used platform. The treatment design in this research is shown in Table 3. Equation (1) is used for the *t*-test meanwhile Eq. (2) is the *N-Gain* equation.

Table 1. Questions blueprint.

	Topics	Question Number
1	Temperature and thermal expansion	1–8
2	Heat	9–14
3	The first law of thermodynamics	15–20
4	Oscillation	21–30, 35
5	Wave	31–34
6	Sound	36–39
7	Stationary Wave	40

$$t = \frac{\overline{X}_1 - \overline{X}_2}{\sqrt{\frac{(n_1-1)S_1^2+(n_2-1)S_2^2}{n_1+n_2-2}\left(\frac{1}{n_1}+\frac{1}{n_2}\right)}} \quad (1)$$

$$\langle g \rangle = \frac{(S_{\text{post}}) - (S_{pre})}{100\% - (S_{pre})}$$

where $\overline{X}_1$ = mean score of the control group mean; $\overline{X}_2$ = mean score of the experimental group; n_1 = the number of samples in the control group; n_2 = the number of samples in the experimental group; S_1^2 = the variance of the control group; S_1^2 = the variance of the experimental group; $\langle g \rangle$ = gain score; S_{post} = posttest score; S_{pre} = pretest score

3 RESULT AND DISCUSSION

The learning process was conducted using Universitas Negeri Malang's LMS Moodle called SIPEJAR. In Figure 1, it shows that there is a pre-class ticket used as discussion material for both groups divided for small group discussion, which was conducted asynchronously and class discussion which was conducted synchronously. Furthermore, there is a web meeting activity for further discussion and reinforcement conducted to strengthen students' understanding based on the conducted discussion. The given problems in the pre-class ticket 9 are chosen as an example of a problem with the solution that can be obtained from the practical simulation and shown in Figure 1.

To analyze the effectiveness of an integrative online platform compared to the LMS-chat learning platform in improving students' problem-solving skills and students' conceptual understanding comprehensivel, a statistical test was conducted. The first step is by conducting normality and homogeneity test toward the control and experimental group. The pre-test score was used to identify students' initial skills meanwhile the post-test score was used to measure students' skills post-treatment. The data of the normality test is shown in Table 5 meanwhile the homogeneity test is shown in Table 6.

Table 5 shows the normality results of pre-test and post-test analysis for control and experimental groups. The value *sig.*>0.05 hence it could be concluded that the data is normally distributed. Furthermore, based on Table 6, the test result shows that the value *sig.*>0.05 hence it can be concluded that each class is homogenous. The descriptive statistics for this data are shown in Table 7.

Based on Table 7, it ca be seen that the mean scores of students' Physics problem-solving skills of control and experimental groups are similar. However, the post-test score for the control group

Table 2. Questionnaire results.

No	Statements	Scales (%)			
		4	3	2	1
1	Students' proficiency and mastery in using a computer makes the learning process easier by using an online platform (LMS Moodle, Practical Simulation, *WhatsApp*, and Web Meeting)	49.4	43.8	4.5	2.2
2	Lecturers' proficiency and mastery in using computers enable students to follow the learning process easily	69.7	25.8	2.2	2.2
3	Learning process using integrative platform makes learning becomes more effective compared to the learning assisted by one of the components	41.6	50.6	6.7	1.1
4	The combination between web meeting, simulation, and conference meeting improve students' learning productivity	36	57.3	5.6	1.1
5	Interaction between students and the lecturer is likely to happen more using the integrative platform	36	55.1	7.9	1.1
6	Integrative online learning encourage effectiveness and students' self-confidence improvement in presenting their ideas based on the given problems	18	67.4	13.5	1.1
7	Online learning using an integrative platform motivates students to study harder and be more active	34.8	58.4	5.6	1.1
8	Learning using an integrative platform succeeded in improving students' physics conceptual understanding comprehensively	16.9	67.4	14.6	1.1
9	Learning using an integrative platform improves students' skills in solving various problems by applying appropriate physics concepts	15.7	62.9	20.2	1.1
10	Learning quality has been successfully improved by using various kinds of integrated media	31.5	60.7	5.6	2.2
11	Online learning has been successfully increasing students participation in the learning process	34.8	57.3	5.6	2.2
Means		34.9	55.2	8.4	1.5

Table 3. The design of quasi-experimental research.

Groups	Pre-treatment	Treatment	Post-treatment
Control	Pre-test about the prerequired course based on conceptual understanding and problem-solving	Asynchronous discussion to solve problems in LMS, *WhatsApp*, synchronous discussion, quiz, and assignment in LMS.	Conceptual understanding and problem–solving-based post-test, effectivity questionnaire for students.
Experimental		Asynchronous discussion to solve problems in LMS, problem–solving-based scientific investigation or using *PhET* simulation, synchronous discussion, and reinforcement using *Google Meet*, further discussion for students in *WhatsApp*, quiz, and assignment using LMS.	

Table 4. *N-Gain* score category.

No	*N-Gain* Score Percentage	Category
1	*N-Gain* > 70	High
2	30 ≤ *N-Gain* ≤ 70	Medium
3	*N-Gain* < 30	Low

Source: (Archambault et al. 2008)

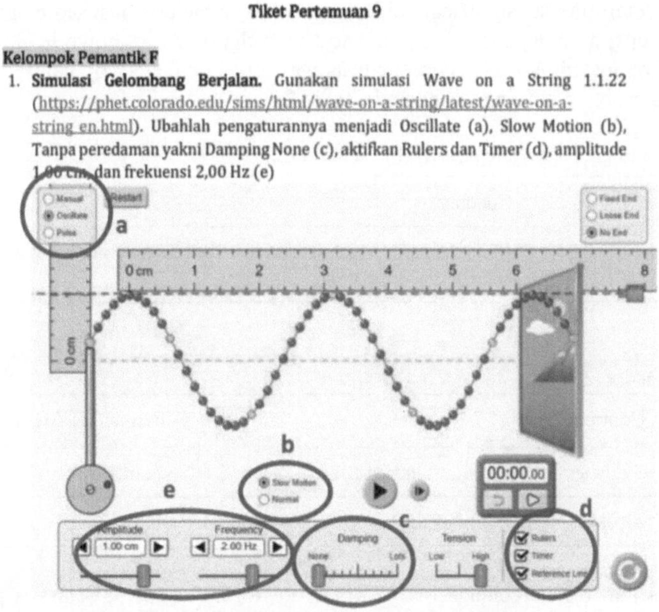

Figure 1. A display example of one of the pre-class tickets that integrates PhET simulation into learning.

Table 5. The results of the problem-solving skill normality test.

		Shapiro-Wilk		
Data Types	Group	Statistics	*df*	*Sig.*
Pre-test	Control	0.962	33	0.287
	Experimental	0.965	33	0.359
Post-test	Control	0.973	33	0.571
	Experimental	0.966	33	0.386

Table 6. The results of students' problem-solving skill homogeneity test.

Data Types	Levine Statistics	*Sig.*
Pre-test	0.903	0.533
Post-test	0.958	0.586

Table 7. Descriptive statistics analysis.

Category	Sample Numbers	Minimum	Maximum	Mean	Deviation Standard
Control: Pre-test	33	50.00	74.00	60.48	6.021
Control: Post-test	33	45.00	73.00	57.55	6.250
Experimental: Pre-test	33	48.00	76.00	60.42	6.942
Experimental: Post-test	33	50.00	78.00	62.61	6.314

where the learning was only conducted using LMS' discussion forum and chat shows decreasing mean. Based on the class evaluation, students felt demotivated by the routines and the lacking of the learning environment and content variation. In contrast, the post-test score of the experimental group where the learning process was using an integrative platform shows an increasing mean.

Furthermore, to determine the significant difference between the post-test score of the control and experimental groups, an independent sample *t*-test with the result is shown in Table 8. After the *t*-test had been conducted, *N-Gain* score analysis was also conducted to analyze the increase of students' physics problem-solving as shown in Table 8.

Table 8. The mean of post-test score post-treatment analysis.

Data Type	*Sig.*		Conclusion
Post-test	0.042	0.05	H_0 is rejected, which means that there is no significant difference between control and experimental groups

Table 9. *N-Gain Score.*

	Control Group		Experimental Group	
	Statistics	Standard Error	Statistics	Standard Error
Mean	−10.42	4.65	3.06	3.82

Table 9 shows that the percentage of *N-Gain* score for experimental class is positive which shows an increase compared to control group which is negative and shows decreasing trend. Hence, learning activity using an integrative platform conducted in the experimental group is more effective compared to learning activity in the control group, which was using only discussion, quiz, and assignment along with chat even though based on Table 3, the increase is relatively low.

A questionnaire to identify students' responses towards learning activity using an integrative platform had been distributed and the result is shown in Table 2. Based on the questionnaire, generally, 90.1% of the respondents agreed with every item on the questionnaire. Thus, most respondents give positive responses to the use of the integrative platform on Fundamental Physics II learning activity. As much as 84.3% of students agreed that the integrative platform could enhance students' physics understanding comprehensively and 78.6% of them agreed that the integrative platform could also improve students' problem-solving skills. Besides, the integrative platform could motivate students to study harder and encourage students to improve their self-confidence to speak in public. Various suggestions were also received from the students such as to maintain the use of the integrative platform for online learning, to divide the students into groups based on their previous achievement, and to combine online learning with offline experiments.

Base on the analysis of the results, LMS Moodle-based online learning integrated with various media such as practical simulation, online conference meeting, and chat application is effective to be used to improve students' problem-solving skills on Fundamental Physics II course. It is in line with a previous study that shows Moodle use could improve students' physics conceptual understanding (Setiawan et al. 2020). Besides, several types of research have also shown that the use of innovative and various media increases students' enthusiasm in the learning activity (Bestari et al. 2020; Huwaidi et al. 2021; Suswanto et al. 2021). It is expected to lessen the impact of online learning experienced by the students including the change of sudden learning style caused by the COVID-19 pandemic (Ratu et al. 2020).

This study has shown that the appropriate choice of online learning strategy could encourage the improvement of students' problem-solving skills even with pandemic limitations. It is in line with a study conducted by several researchers that shows students' problem-solving skills could be developed even through online learning (Bixler & Bixler 2007; Demiraslan Çevik et al. 2015; Tawfik et al. 2014). Students' enthusiasm, motivation, and activity improvement in online class could be supported by the fact that students learn in their comfort zone where they are free to explore since they are not limited by space and manners such that in offline learning (Al-Salman 2009; Bonk & Graham 2012; Bonk & King 2012).

4 CONCLUSIONS

Based on the conducted study, it could be concluded that learning activity using the integrative online platform could improve students' problem-solving skills effectively compared to the one that only uses LMS features and online chat applications. This study has also demonstrated the use of the online platform for physics learning. However, this study was limited only to the Fundamental Physics course for science students hence its potential to be used in other courses by integrating other media is still widely open.

REFERENCES

Abidah, A., Hidaayatullaah, H.N., Simamora, R.M., Fehabutar, D., Mutakinati, L., 2020. The impact of covid-19 to indonesian education and its relation to the philosophy of "merdeka belajar." Stud. Philos. Sci. Educ. 1, 38–49.

Al-Salman, S.M., 2009. The role of the asynchronous discussion forum in online communication.J. Instr. Deliv. Syst. 23, 8–13.

Archambault, J., Burch, T., Crofton, M., McClure, A., 2008. The Effects of Developing Kinematics Concept Graphically Prior to Introducing Algebraic Problem Solving Techniques. Action Research Required for the Master of Natural Science Degree with Concentration in Physics. Arizona State University.

Bestari, A.C.Y., Faiza, D., Mayekti, M.H., 2020. Instagram Caption As Online Learning Media On The Subject Of Extended Writing During Pandemic Of Covid-19. Surakarta English …3, 9–21.

Bixler, B.A., Bixler, B.A., 2007. INTRINSIC MOTIVATION IN AN ONLINE LEARNING ENVIRONMENT by A Thesis in.

Bonk, C.J., Graham, C.R., 2012. The handbook of blended learning: Global perspectives, local designs. John Wiley & Sons.

Bonk, C.J., King, K.S., 2012. Models of asynchronous computer conferencing for collaborative learning in large college classes, in: Electronic Collaborators. Routledge, pp. 193–218.

Cardwell, V.B., 2005. Literacy: What level for food, land, natural resources, and environment? J. Nat. Resour. life Sci. Educ. 34, 112–117.

Demiraslan Çevik, Y., Ha°laman, T., Çelik, S., 2015. The effect of peer assessment on problem solving skills of prospective teachers supported by online learning activities. Stud. Educ. Eval. 44, 23–35. https://doi.org/https://doi.org/10.1016/j.stueduc.2014.12.002

Goldschmidt, K., 2020. The COVID-19 pandemic: Technology use to support the wellbeing of children. J. Pediatr. Nurs. 53, 88.

Gultepe, N., Kilic, Z., 2015. Effect of Scientific Argumentation on the Development of Scientific Process Skills in the Context of Teaching Chemistry. Int. J. Environ. Sci. Educ. 10, 111–132.

Huwaidi, F., Nandiyanto, A.B.D., ..., 2021. The urgency of online learning media during the Covid-19 pandemic at the vocational school in Indonesia. Indones. J. …1, 35–40.

Liu, X., 2009. Beyond science literacy: Science and the public. Int. J. Environ. Sci. Educ. 4, 301–311.

Ratu, D., Uswatun, A., Pramudibyanto, H., 2020. Pendidikan Dalam Masa Pandemi Covid-19. J. Sinestesia 10, 41–48.

Setiawan, B., Juniarso, T., Fanani, A., Iasha, V., 2020. Pembelajaran Online Di Masa Pandemi Covid-19: Pengaruhnya Terhadap Pemahanan Konsep Fisika Mahasiswa. J. Pendidik. Dasar 11, 230–236.

Sukmana, M., Aminuddin, M., Nopriyanto, D., 2020. Indonesian government response in COVID-19 disaster prevention. East African Sch. J. Med. Sci. 3, 81–86.

Sun, L., Tang, Y., Zuo, W., 2020. Coronavirus pushes education online. Nat. Mater. 19, 687.

Suswanto, B., Sulaiman, A.I., Sugito, T., Weningsih, S., Sabiq, A., Kuncoro, B., 2021. Designing Online Learning Evaluation in Times of Covid-19 Pandemic. Int. Educ. Res. 4, p18. https://doi.org/10.30560/ier.v4n1p18

Tawfik, A.A., Sánchez, L., Saparova, D., 2014. The effects of case libraries in supporting collaborative problem-solving in an online learning environment. Technol. Knowl. Learn. 19, 337–358. https://doi.org/10.1007/s10758-014-9230-8

Treagust, D.F., Duit, R., Fischer, H.E., 2017. Multiple representations in physics education. Springer.

Improving Assessment and Evaluation Strategies on Online Learning – Wonorahardjo, Karmina & Habiddin (eds)

Improving vocational education quality through life-based learning

Yoto & P.A.N. Mawangi*
Universitas Negeri Malang, Indonesia

A.N. Pramudhita
State Polytechnic of Malang, Indonesia

ABSTRACT: Vocational graduates supposedly have the competencies following the purpose of vocational education: preparing medium-skilled labour in a particular area that corresponds with the industrial needs. In the 21st century, businesses and industries tend to demand skilled labours in digital literacy and human literacy. Thus, vocational education should be able to innovate to respond to the challenge and opportunity from the 21st-century implication through revitalizing the vocational education, involving: learning system, education units, learners, educators, and educational staffs. A suitable learning system for the 21st century is a life-based learning model. Life-based learning in vocational education could be performed through industrial practice, industrial class, and teaching factory activities. These three approaches aimed to bring the learning activities closer with real-life work at the workplace through field practices. With life-based learning at the real workplace, students would develop a discipline and work culture habits as provisions after finishing the vocational education.

Keywords: productive teacher competence, learning, skill competency

1 INTRODUCTION

The government's policy to improve Indonesian human resources is by improving education quality. Following the changes in the educational and business environment in the 21st century, professionalism in all aspects, including in the educational area, is required. A tight competition and limited jobs condition decreased the opportunities for graduates to be accepted. The Head of East Java Educational Authorities stated that there was 64.7% of graduates' absorbance in the work field [1]. This condition was far from the planned target by the Ministry of Education and Culture that hoped for 75% of graduates' absorbance in the work field [2]. The low chance of vocational education graduates to be accepted in business/industry, as mentioned above, influences the increasing unemployment rate for vocational graduates in Indonesia.

The identifying factor that was significantly affecting the condition is the quantity of vocational education in Indonesia, reaching 12,659 [3]; however, the amounts are not in line with the learning process quality and the produced output quality. The majority of vocational education nowadays tends to fill the supply-driven basis, only stands to supply graduates, but forgets the demand-driven basis, or not adjusting to each area's requirement. The consequence is a lack of quality alignment from the graduates of each school that follow the potency or chance in each region. The impact is a lack of hard skill and soft skill from the graduates.

A survey of 460 companies in Java, Sumatera, Kalimantan, and Sulawesi showed that the level of competency from vocational education did not correspond with the company needs [4]. It showed

*Corresponding Author

 DOI 10.1201/9781003261346-15

that the pragmatism philosophy base of "theory is right if it works", that vocational education stands/exists to fulfil the needs in society, did not well-realized [5].

From the above description, the primary problem was the quality of vocational graduates in Indonesia who were not ready to work. The solution was through the 21st-century educational concept, such as soft skill development for vocational education students that covers critical thinking, creative thinking, collaborative, and communicative [6]. Following that, the Director-General of Science and Technology and the Director-General of Higher Education [7] in 2018 National Work Meeting of the Ministry of Research and Higher Education stated that there are human resource skills that Indonesians must fulfil (including graduates) to face the 21st-century learning that consists of four competencies: (1) critical thinking, (2) creativity, (3) communication, and (4) collaboration. These 21st-century skills would benefit the vocational students in developing working readiness, improving confidence, performance, independent work, and capabilities. Students with those skills would be able to use, develop the skilled competency after graduating. A learning method that could transform human resources into a competent one is life-based learning [8].

2 RESULTS AND DISCUSSION

2.1 *21st-century learning quality improvement*

According to Trilling and Fadel [9], 21st-century learning is oriented to the digital lifestyle, thinking tool, learning research, and knowledge procedure. Three out of the four 21st-century learning orientations are closely related to vocational education. They are knowledge procedure, strengthening the thinking tool, and digital lifestyle. Knowledge procedure is the ability to collaborate in a team with different location and tools. Strengthening the thinking tool is the skill to use the technology, digital tool, and service. The digital lifestyle is the capability to use and adjust in the digital era [9].

The world economic forum predicted that the 21st-century skills structure would change. In 2015, the skill structures were: (1) complex problem solving, (2) cooperation with others, (3) people management, (4) critical thinking, (5) negotiation, (6) quality control, (7) service orientation, (8) decision assessment and making, (9) active listening, and (10) creativity. Meanwhile, in 2020, they change into (1) complex problem-solving, (2) critical thinking, (3) creativity, (4) people management, (5) cooperation with others, (6) emotional intelligence, (7) decision assessment and making, (8) service orientation, (9) negotiation, and (10) cognitive flexibility [10].

All form of skills in the 21st-century should be integrated into all educational component, starts from the learning system, educational units, learners, up to the educators and educational staffs. Besides, there needs support from society to improve the excellent image of vocational education so that graduates feel safe as skilled workers because of the support and recognition from society. Vocational education can be provided and facilitated by society and the government to quickly prepare and transform an individual in fulfilling the workplace demands [11].

The 21st-century learning content should be susceptible to time dynamics. The learning contents that are expected to meet the 21st-century skills are (1) innovation learning and skill consist of mastering various knowledge and skills, learning and innovation, critical thinking and problem-solving, communication and collaboration, and creativity and innovation; (2) digital literacy skill that covers information literacy, media literacy, and ICT literacy; and (3) career and life skills involving flexibility and adaptability, initiative, social and cultural interaction, productivity and accountability, and leadership and responsibility [9].

The elements that interact in the chronosystem must integrate their focuses on physical, digital, and biological. Elements in vocational education, particularly vocational education as a part of chronosystem, must strengthen the new literacy movement (digital literacy, technology literacy, and human literacy). The reinforcement is performed to add values and competitiveness from vocational education graduates. Figure 2 displays the interaction and integration between the elements.

Figure 2 shows the integration from all vocational education components that should be mediated by the vocational education system because vocational education has a vast interest to mediate all

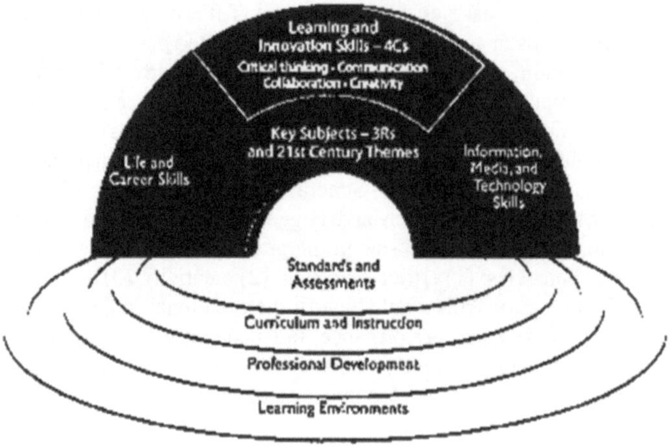

Figure 1. Core subject 21st-century skills.

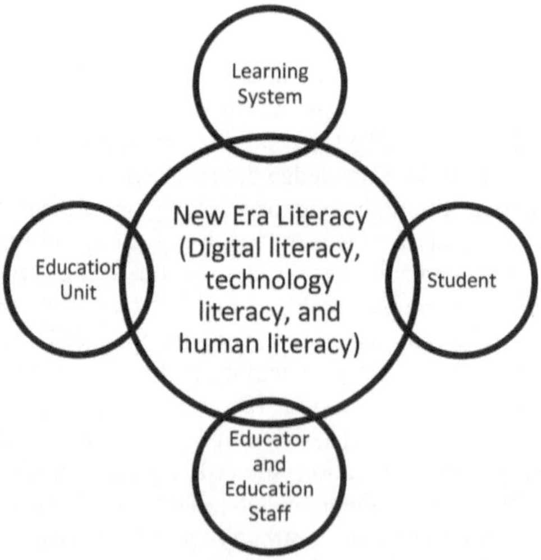

Figure 2. Chronosystem of vocational education in the Industrial 4.0 Era.

elements that aim to improve the quality of learning system, educational system, learners, and educators and educational staffs to create competitive graduates.

2.2 *Life-based learning*

Life-based learning is a model that was developed by Maret Staron who worked as a teacher, a consultant in TAFE NSW International Centre for Vocational Education and Training (VET), and the director of TAFE NSW Workforce Development. The focus from life-based learning is capabilities development (ability and will) in the era that demands mastery of knowledge to contribute to social welfare and happiness. According to Sudira [12], knowledge capability is measured from the usefulness of the developed knowledge (widyaguna) in developing prosperity and happiness together. The knowledge that benefits the well-being and happiness of all human beings should be developed in the field of science. At the same time, knowledge which does not benefit well-being and happiness must be obsolete in the warehouse.

Useful knowledge is a knowledge that can be practised, the knowledge that builds life habits that benefit others. Life-based learning cannot be separated from the opinion of Richard Slaughter that was published in 2005 paper in which it reviewed and compared various paradigms that could be used and developed to be relevant in the educational area. Based on said published paper, views and paradigms can be described into five: environmental, neo-humanists/multicultural/de-colonizing, spiritual, futures, integral [13].

Staron [14] stated that the process of obtaining knowledge and skills of understanding the nature of life, skilled in solving life problems and living a balanced and harmonious life. Life-based learning explores the concept that learning from life if real learning; therefore, human must exists in society. Life-based learning, as seen from Indonesia educational perspective, with education as its primary, the learning process aims to form a complete human because the essence of education is to humanize human.

2.3 *Life-based learning implementation*

Life-based learning (LBL) in vocational education learning is applied to industrial practice, industrial class, and teaching factory activity. All three give practical experience and in-field activity to shape adequate behaviour, knowledge, and skill.

Industrial practice is the activity that is performed by students after taking at minimum five semesters learning activity in vocational high school (VHS) and have completed essential practice learning in the vocational school. Industrial practice is conducted outside of the school, in the industry partner of the school with mutual benefit. Industrial practice is life-based learning that the students performed in the industry to carry out practical work practices in the field. Industrial practice is conducted with instructor guidance structurally and intensively to shape a discipline, honest, responsible, and have an industrial culture.

Industrial class is a learning activity that is conducted together by the vocational school and industry [15]. From recruitment, learning implementation, evaluation, and industrial practice are collaboratively done between vocational school and the industry. Thus, the industry is a partner in designing vocational education so that the graduates have the demanded competency. Therefore, graduates from the industrial class will be recruited by the industry partner.

Teaching factory is a learning activity that is performed by students in the school where the learning activity is a practicum or a work that the results can be utilized by society and/or industry [16]. Thus, the machines/devices in teaching factory are standardized machines/devices because the product must follow the industry standards and directly used by society/industry. From here, students learn how to make products following the standard and provision of the industry; therefore, shaping the industrial culture in the activity.

3 CONCLUSION

Vocational graduates should have the competencies that met the purpose of vocational education or obtained medium level skill in a particular area following the demanded skill from the business and industry.

In the 21st-century, business and industry tended to require workers who were skilled in digital literacy, technology literacy, and human literacy. Thus, vocational education should be able to innovate and create in responding the challenge and opportunity from the 21st-century implication through the revitalization of vocational education that consisted of the learning system, educational units, learners, and educators and educational staffs. A suitable learning system in the 21st-century was the life-based learning model.

The application of life-based learning in vocational education could be seen in three examples of learning activities: industrial practice, industrial class, and teaching factory. All learnings influenced students to improve their working skills, shaping discipline, honesty, responsibility, and working culture in performing a job.

4 SUGGESTION

Because the advantage of life-based learning for vocational students was quite huge, it is expected for the vocational education management to perform habituation and work culture for the students so that when they graduate, they would have strong skill and knowledge to compete in the workplace and/or create employment for themselves and others. Vocational education management is expected to improve cooperation with business/industry in performing industrial practice, industrial class, and teaching factory to carry out life-based learning.

REFERENCES

Hasanah N U Serapan Kerja Lulusan SMK di Jatim Sebanyak 64,7 Persen per Tahun Surya Malang

Apriliyadi Kemendikbud Targetkan Capaian SMK pada 2020 Dir. SMK

Anon 2016 Statistik Sekolah Menengah Kejuruan (SMK) tahun 2015/2016

Brodjonegoro S S 2016 Revitalisasi Pendidikan Kejuruan Kompas 6

Widarto P and Widodo N 2012 Pengembangan model pembelajaran soft skills dan hard skills untuk siswa SMK Cakrawala Pendidik. 31 409–23

Murti K E 2013 Pendidikan Abad 21 Dan Implementasinya Pada Pembelajaran Di Sekolah Menengah Kejuruan (SMK) Untuk Paket Keahlian Desain Interior Artik. Kurikulum 1–23

Anon 2018 Menyiapkan Sumber Daya IPTEK dan DIKTI dalam Revolusi Industri 4.0 (Rencana Induk Pengembangan Sumber Daya Manusia Indonesia)

Staron M, Jasinski M and Weatherley R 2006 Life Based Learning: A Strength Based Approach for Capacity Development in Vocational and Technical Education: A Report on the Research Project Designing Professional Development for the Knowledge Era (Tafe NSW)

Trilling B and Fadel C 2009 21st century skills: Learning for life in our times (John Wiley & Sons)

Irianto D 2017 Industry 4.0; The Challenges of Tomorrow

Murgor T K 2013 Relationship between Technical and Vocational Acquired Skills and Skills Required in Job Market; Evidence from TVET Institutions, Uasin Gishu County, Kenya J. Educ. Pract. 4 77–83

Sudira P 2014 Life-Based Learning dalam Pendidikan Teknologi dan Vokasional Tanpa Tahun See Httpstaffnew Uny Ac Idupload131655274penelitianlife-Basedlearning Pdf 2

Slaughter R 2005 Emerging paradigms in the Knowledge Era

Staron M Learning to be professional / Life based learning

Yoto Y, Marsono M, Irdianto W and Basuki B 2019 Industrial Class With Work Based Learning Approach As Alternative To Increase Educational Quality In Vocational High School 2nd International Conference on Vocational Education and Training (ICOVET 2018) (Atlantis Press)

Yoto and Marsono 2019 Implementation of Work-Based Learning at Teaching Factory in Vocational education International Conference on Vocational Education and Training (ICOVET 2019)

M. P. Brown and K. Austin, *The New Physique* (Publisher Name, Publisher City, 2005), pp. 25–30.

M. P. Brown and K. Austin, Appl. Phys. Letters 85, 2503–2504 (2004).

R. T. Wang, "Title of Chapter," in *Classic Physiques*, edited by R. B. Hamil (Publisher Name, Publisher City, 1999), pp. 212–213.

C. D. Smith and E. F. Jones, "Load-cycling in cubic press," in *Shock Compression of Condensed Matter-2001*, AIP Conference Proceedings 620, edited by M. D. Furnish *et al.* (AIP Publishing, Melville, NY, 2002), pp. 651–654.

B. R. Jackson and T. Pitman, U.S. Patent No. 6,345,224 (8 July 2004)

D. L. Davids, "Recovery effects in binary aluminum alloys," Ph.D. thesis, Harvard University, 1998.

R. C. Mikkelson (private communication)

Improving Assessment and Evaluation Strategies on Online Learning – Wonorahardjo, Karmina & Habiddin (eds)

Investigating the memory retention in extension concept mapping

D.D. Prasetya*
Universitas Negeri Malang, Indonesia

A. Pinandito
Universitas Brawijaya, Indonesia

Y. Hayashi & T. Hirashima
Hiroshima University, Japan

ABSTRACT: Extension concept mapping has been found to engage learners in a wide range of learning activities while also improving knowledge structures. Previous research compared two ways to extension concept mapping: Extended Scratch-Build and Extended Kit-Build. Students who utilized the Extended Kit-Build had higher understanding test scores, according to the findings. However, the initial research did not look into how the two approaches affect students' memory retention when it comes to recalling what they've learned. This study investigates the effect of memory retention on the two comparable methods. Extended Scratch-Build extends an open-ended concept map way, while Extended Kit-Build extends a closed-ended Kit-Build with an open-ended fashion. Kit-Build is a recomposition map that encourages students to recompose a kit according to the knowledge objective, using the teacher's maps as a guide. Fifty-five university students who enrolled in the Database 1 course were involved as participants. Since the participants were homogeneous, group assignment was done randomly. The experimental group used the Extended Kit-Build, and the control group used the Extended Scratch-Build. A delayed comprehension exam was administered two weeks following the initial test to assess group performance. The results show that the experimental group's memory retention was much better than the control group. In particular, experimental achievement differed significantly for questions in the kit, but for questions not in the kit, although achievement was constant, no significant discrepancy was glimpsed.

Keywords: concept map, delayed test, extension concept mapping, memory retention

1 INTRODUCTION

Concept maps are graphical representations of individual conceptual knowledge. Therefore, they are widely used for learning, teaching, and evaluating students at various stages of teaching. It has been proven that concept maps have a favorable influence on learning and improve learners' comprehension. It can be used individually or collaboratively in traditional face-to-face and online class environments. Several recent studies have also demonstrated that the concept map approach is suitable for use in blended learning and fully online education in the COVID-19 pandemic situation (Hornink & Costa 2021; ;inandito et al. 2021; Prasetya et al. 2020a;. The concept map engages learners more actively and is able to improve student achievement better than usual online learning.

Another positive impact of using concept maps in learning is to encourage students' long-term memory retention. Kim and Olaciregui (2008) investigated two groups of students using a folder-based and concept map-based information system. The results reported that the concept

*Corresponding Author

DOI 10.1201/9781003261346-16

map approach encourages students to get higher comprehension test scores in the fastest time and significantly achieve delayed-test scores. In a more particular study, Alkhateeb et al. (2016) revealed that the recomposition of the concept map through the Kit-Build (KB) framework helps the learners retain information for more extended periods. Similar research was conducted by Andoko et al. (2020), which compared the KB map with and without the source connection and traditional learning. The results emphasized that the KB approach facilitates learners to recall information that has been obtained better than other methods. While there has been researched on the effects of concept mapping on memory retention, there have been few studies on the impact of extension concept mapping.

Extension mapping is a strategy for expanding a concept map that already exists by adding relevant concepts and relationships. As a result, there will be at least two types of concept maps in the extension concept mapping activity: original maps and additional maps (Prasetya et al. 2021). The original map is an initial map that refers to the original material, whereas the additional map is a more advanced concept map based on the additional material. Extension concept mapping offers commensurate activities to generate increased understanding and knowledge structure. Through two phases of concept mapping, learners could review their initial ideas, uncover missing relationships, add new components, and produce better knowledge structures (Foley et al. 2018; Schwendimann & Linn 2016). The design of the extension concept mapping has the opportunity to reduce the cognitive load caused by the complexity of the map content.

Previous studies introduced two comparable extension mapping designs: Extended Scratch-Build (ESB) and Extended Kit-Build (EKB) map. The ESB builds the initial map using an open-ended approach and then extends it to create the additional map. Open-ended allows learners to add concept map components according to their understanding, while closed-ended provides defined components to be reconstructed. Slightly different, the EKB extends a recomposition and closed-ended KB framework by allowing learners to add new components. Recomposition is a vital learning practice that enables students to comprehend the teacher's thinking by reconstructing the teacher's map. KB demonstrates that recomposition concept mapping allows students to adapt previously acquired abilities to new situations and circumstances (Hirashima 2019). In practice, KB requires that a learner rebuilds a concept map using a kit (a collection of nodes and links) created by disassembling an expert's map (Hirashima et al. 2015).

Initial studies revealed that the ESB map motivates learners to improve their comprehension scores, including immediate-test and delayed-test (Prasetya et al. 2020b). Another study compared the effects of ESB and EKB on students' immediate-test scores and map size (Prasetya et al. 2021). The results indicated that the achievement of students using EKB was superior to ESB. Although the extension concept mapping has a potential strategy, no further discussion has analyzed its effect on memory retention. Retention is one of the most significant educational goals that demonstrate meaningful learning (Mayer 2002). Retention refers to the ability to recall information in the same way that it was delivered during education. In fact, when knowledge is applied to a more complex activity, remembering it is critical for effective learning and problem-solving. This study hypothesizes that the EKB approach that utilizes the KB kits facilitates memory retention more than ESB.

2 METHODS

2.1 *Measurement*

The present work was done in the Database 1 subject, and the teacher covered the topic of Relational Database using Indonesian. In the control and experimental groups, immediate and delayed tests were used to assess students' comprehension of the Database content. The immediate test was designed to assess students' performance immediately following treatment, whereas the delayed test was used to assess students' success after a period of time had passed. To confirm the students' knowledge, eight multiple-choice questions (4 connected to the original content and four pertaining to additional material) were employed. The students' grasp of the immediate-test and delayed-test was assessed using the same questions but in a different sequence.

2.2 *Setup procedures*

ESB and EKB are extension concept mapping tools that divide learning activities into two parts: Phase 1 and Phase 2. These two phases are carried out in one day of learning activities. In Phase 1, learners are asked to create an original map, while in Phase 2, they are asked to produce an additional map. Figure 1 depicts the experimental flow in this investigation. Students in both groups used the same experimental procedure.

Participants were given an introduction to idea maps and how to utilize them before the experiment began at a previous course meeting. Participants utilized a straightforward case study that was designed to help them adjust to a new system. They were allowed to use the EKB concept map tool in the classroom and outside the classroom when the lecture was finished. Furthermore, after the participants understood how the system works and is used, they were involved in an experiment to identify their system's perceptions. During the concept mapping, the lecturer acts as a facilitator.

The exercise begins with a pre-test to assess the starting ability before receiving instruction in Phase 1. In addition, for 25 minutes, each group got the original part material from the lecturer. The students in the control group were then asked to use an open-ended idea map to convey their knowledge of the original content. In the experimental group, students were requested to recompose the concept map using the KB approach. The concept map creation in both groups was carried out for 15 minutes, and the teacher acted as a facilitator. During the map construction, students could read the material handouts that have been distributed.

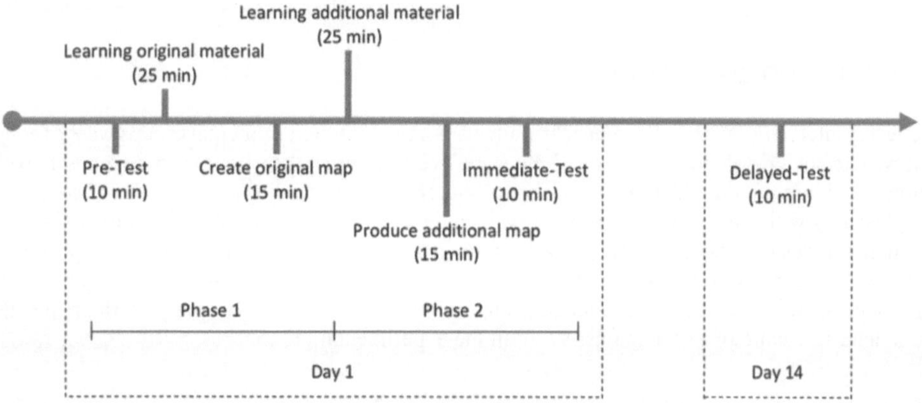

Figure 1. Experiment flow.

Phase 2 activities followed the same schedule as Phase 1. The teacher continued to provide the additional part content in the same manner as previously, with the same strategy and allocation. Additionally, both sets of students were encouraged to expand their idea maps using the same procedure, which was open-ended. Thus, they were allowed to add new components and link them to the previous original map. The concept map expansion activity was also carried out for 15 minutes. Finally, both groups were given an immediate test to assess their comprehension of the Database material.

Delayed-test was intended to measure the ability of learners to remember the knowledge that has been obtained previously. The test was carried out 14 days after they received the immediate test. The questions in the delayed test were the same as in the immediate test but in randomized order of questions. Thus, this approach could avoid learners who only remember the sequence. Delayed-test was given without informing the learners beforehand. Therefore, learners were not prepared to learn in advance and relied on their memory retention related to the knowledge that had been obtained.

2.3 *Participants characteristics*

The research included second-year students from an Indonesian public university's Department of Electrical Engineering. A total of 55 students from two regular Informatics Engineering courses took part in the experiment. Referring to the initial questionnaire, all participants had never used a concept map before. After obtaining the condition that the two classes were homogeneous, group assignments were carried out randomly. The control group consisted of 27 students from class A, whereas the experimental group consisted of 28 students from class B. Based on the characteristics of the participants, it could be said that they have the same characteristics.

2.4 *Data analysis*

The dataset for this study was collected from the participants divided into two groups: control and experimental. The normalcy distribution and variance homogeneity were investigated. The results indicated that both participants were homogeneous, while normality was not found. Therefore, a non-parametric technique was used to do the statistical analysis. The significance of the difference in performance between the control and experimental groups was determined using the Mann-Whitney U test. The Wilcoxon signed-order test was used to evaluate the differences between the control and experimental groups' pre-test and immediate-test results. The correlation coefficient was also examined using Pearson's r as the effect size (ES) measure. Statistical significance was defined as a p-value of less than 0.05.

3 RESULT AND DISCUSSION

Pre-test results were used to guarantee that students in control and experimental groups were of equal ability. The students in both groups had the same pre-test comprehension accomplishment levels before receiving the intervention, according to the preliminary findings. The students' successes after learning with the concept maps were evaluated using an immediate test of comprehension of instructional design. The descriptive data for both groups' immediate-test outcomes are presented in Table 1. According to the data, the two groups' different treatments resulted in higher accomplishment. Despite this, the experimental group outperformed the control group, with an average immediate-test score of 81.02 against 87.50 in the experimental group.

Table 1. Descriptive statistics of the immediate-test scores for both groups.

Group	N	Minimum	Maximum	Median	Mean	SD
Control group	27	62.50	100	87.50	81.02	10.612
Experimental group	28	62.50	100	87.50	87.50	10.758

The Mann-Whitney U statistical test was used to compare the immediate-test results of the control and experimental groups after the intervention. The findings stated a difference between the groups' achievements ($Z = -2.099$; $p = 0.036$). Based on a significance level of 0.05, the results showed a statistically significant difference. The Pearson's r results were 0.283, which indicated a small effect size.

The delayed-test of understanding in instructional design was used to assess the students' performance after several days of the extended idea mapping exercise. Table 2 shows the representative statistics of the delayed-test results for both groups. The results show that after a delay of 14 days, students in both groups experienced a decrease in achievement. The average delayed-test score in the control group was 69.91, while the experimental group was 78.57.

Table 2. Descriptive statistics of the delayed-test scores for both groups.

Group	N	Minimum	Maximum	Median	Mean	SD
Control group	27	25	87.50	75.00	69.91	13.98
Experimental group	28	62.50	100	75.00	78.57	11.71

Figure 2 displays the visual comparison between the pre-test, immediate-test, and delayed test scores for the control group that used the ESB method and the experimental group that utilized the EKB approach. Students in both groups had the same initial skills as indicated by the achievement of pre-test scores. However, after receiving a different intervention in Phase 1 and continuing with the same intervention in Phase 2, the experimental group had a higher immediate-test score than the control group. Furthermore, after a delay of 14 days, although both groups experienced decreased performance, students in the experimental group achieved a higher score.

For both groups, the Wilcoxon signed-order test was employed to compare the immediate-test and delayed-test scores pairwise. This comparison value represents the level of improvement in each group's achievement. The Wilcoxon signed-rank test is similar to the paired t-test, except it's used for data that isn't normally distributed, as the data in this research.

The analysis results in the control group indicated a significant difference between the immediate-test and delayed-test scores ($Z = 2{,}930$; $p = 0.003 < 0.05$). Cohen's Pearson's r correlation was 0.399, which indicated a medium effect size. The same condition also occurred in the experimental group using the EKB concept mapping approach. There was a significant discrepancy between the immediate-test and delayed-test attaintment ($Z = 3.201$; $p = 0.001 < 0.05$), with Pearson's r of 0.428, showing a medium effect size. Pairwise test results reported that memory retention in both groups decreased equally after 14 days.

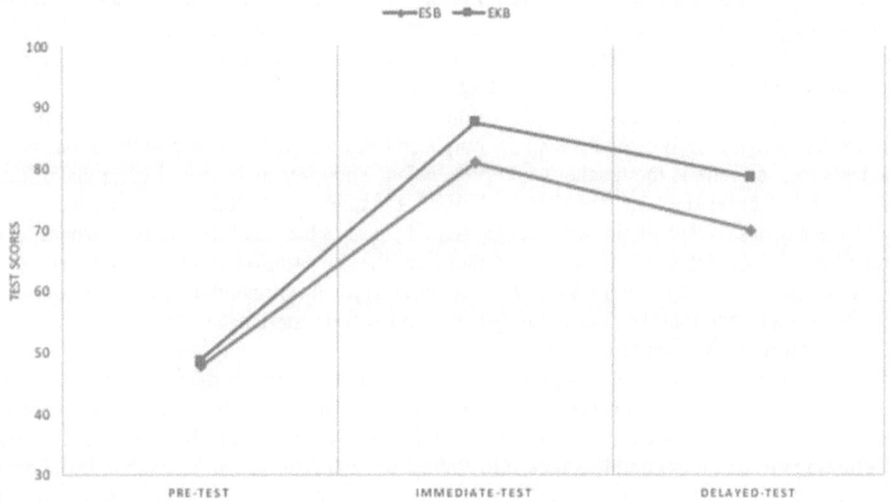

Figure 2. Pre-test, post-test, and delayed-test scores for both groups.

The Mann-Whitney U statistical test was used to further analyze the delayed-test score comparison. According to the findings, there was a substantial difference in knowledge between the control and experimental groups ($Z = -2.228$; $p = 0.026 < 0.05$). Pearson's $r = -0.301$ In terms of post-test accomplishment, this suggests a medium impact size. The control group's average rank was 69,907, whereas the experimental group was 78,571. Despite the fact that both groups' delayed-test scores declined, the experimental group's achievement was higher than the control group's.

The delayed-test questions were divided into two categories to study further the students' memory retention: questions in original content and questions material. The teacher has previously marked the questions relevant to the first and second materials. In both groups, Table 3 provides the descriptive statistics of the delayed-test results in original and added material. On these two categories of questions, students in the experimental group outperformed those in the control group.

Table 3. Descriptive statistics of the question in original and additional materials for both groups.

Question	Group	N	Minimum	Maximum	Median	Mean	SD
In original material	Control group	27	50	100	75.00	79.63	18.39
	Experimental group	28	50	100	100	90.18	14.17
In additional material	Control group	27	0	100	50.00	60.19	19.93
	Experimental group	28	25	100	75.00	66.96	23.62

On two separate question circumstances, the Mann-Whitney U test was employed to determine the difference in accomplishment between the two groups. There was a considerable difference between the control and experimental groups on the questions in the original material ($Z = -2.234$; $p = 0.026 < 0.05$), with Pearson's r of -0.301, showing a medium effect size. The experimental group had a mean rank of 85.19, whereas the control group had a mean rank of 79.630. This study discovered no significant differences between the control and experimental groups on the questions in the additional material ($Z = -1.714$; $p = 0.087 > 0.05$). Pearson's r was -0.139, which implies a modest impact size based on the Z value. Although there was no statistically significant difference, the experimental group's delayed-test score for questions in additional material was greater than the control group.

According to the current findings, pupils who learned using EKB idea mapping outscored those who learned using the ESB technique in terms of memory retention. Furthermore, although in Phase 1, the two groups used different approaches, in Phase 2, both used the same technique. There are several possible explanations for this finding, as discussed below.

First, experimental group students employed the KB framework that offers a recomposition concept mapping approach. KB provides a kit decomposed from the teacher's map to be reconstructed by students to comprehend the teacher's comprehension. The kit is an essential component of the KB system (Hirashima et al. 2015; Pailai et al. 2017; Pinandito et al. 2021), allowing individuals to build their own concept of the knowledge goal. In particular, kits' use helps learners retain and recall previously obtained information better than the open-ended strategy (Alkhateeb et al. 2016). The present finding is consistent with previous studies (Hirashima 2019; Prasetya et al. 2021), which suggested that the map rebuilding method encourages learners to achieve maximum comprehension scores in a particular subject.

Second, the extension concept mapping activity is an appropriate approach to actively involve learners and improve their performance. Map expansion design efficiently organizes design work and establishes a solid knowledge base (Foley et al. 2018). Reviewing original ideas and connections, eliciting missing ideas and linkages, and adding new concepts are all tasks that the extended concept map performs at each level of the knowledge-integration process (Schwendimann & Linn 2016). Although both groups allow learners to expand their concept maps, EKB has prepared a solid original map structure. Thus, this technique makes it easier for learners to expand the concept map and increase their retention.

Third, the EKB method offers a suitable design to facilitate enhanced meaningful learning. Through the two-phase map construction, EKB realizes improved meaningful learning. As stated by Meyer (2002), one of the important foundations of meaningful learning is retention. The current findings revealed that the experimental group's average delayed-test scores were higher than the control group.

4 CONCLUSION

The purpose of this study was to see how extended idea mapping exercises affected memory recall. Extension Scratch-Build (ESB) and Ex-tended Kit-Build (EKB) maps were compared and evaluated further. The results emphasized that students who used the EKB approach outperformed those who utilized ESB in terms of retention evaluated after 14 days. The EKB group received a better score for the sorts of questions in the original map and questions in the supplementary map. In the circumstances of the questions in the supplementary map, however, there was no significant difference between the two groups.

Although this study was successful in answering the hypothesis, there are a few limitations that should be considered in future research. Important considerations are the number of participants in the study and the number of questions used to assess students' recall. To achieve more reliable results, future research should involve more participants and questions related to learning materials.

REFERENCES

Alkhateeb, M., Hayashi, Y., Rajab, T., & Hirashima, T. 2016. Experimental use of kit-build concept map system to support reading comprehension of EFL in comparing with selective underlining strategy. *International Journal of Advanced Computer Science and Applications*, 7(4).

Andoko, B. S., Hayashi, Y., Hirashima, T., & Asri, A. N. 2020. Improving English reading for EFL readers with reviewing kit-build concept map. *Research and Practice in Technology Enhanced Learning*, 15(1), 1–19.

Foley, D., Charron, F., & Plante, J. S. 2018. Potential of the Cogex Software Platform to Replace Logbooks *in Capstone Design Projects. Advances in Engineering Education*, 6(3), n3.

Hirashima, T., Yamasaki, K., Fukuda, H., & Funaoi, H. 2015. Framework of kit-build concept map for automatic diagnosis and its preliminary use. *Research and Practice in Technology Enhanced Learning*, 10(1), 17.

Hirashima, T. 2019. Reconstructional concept map: automatic Assessment and reciprocal reconstruction. *International Journal of Innovation, Creativity and Change*, 5, 669–682.

Hornink, G., & Costa, M. J. 2021. Constructing online concept maps in CMap Cloud collaboratively: Connecting pathways in case scenarios. *Biochemistry and Molecular Biology Education*, 49(1), 29–31.

Kim, P., & Olaciregui, C. 2008. The effects of a concept map-based information display in an electronic portfolio system on information processing and retention in a fifth-grade science class covering the Earth's atmosphere. *British Journal of Educational Technology*, 39(4), 700–714.

Mayer, R. 2002. Rote versus meaningful learning. Theory Into Practice. 41(4), 226–232.

Pailai, J., Wunnasri, W., Yoshida, K., Hayashi, Y., & Hirashima, T. 2017. The practical use of Kit-Build concept map on formative assessment. *Research and Practice in Technology Enhanced Learning*, 12(1), 20.

Pinandito, A., Prasetya, D. D., Hayashi, Y., & Hirashima, T. 2021. Design and development of semi-automatic concept map authoring support tool. *Research and Practice in Technology Enhanced Learning*, 16(1), 1–19.

Prasetya, D. D., Pinandito, A., Hayashi, Y., & Hirashima, T. 2020a. The Performance of Extended Scratch-Build Concept Mapping Tool in Blended Learning. *In 2020 4th International Conference on Vocational Education and Training* (ICOVET) (pp. 345–349). IEEE.

Prasetya, D. D., Hirashima, T., Hayashi, Y. 2020b. Study on Extended Scratch-Build Concept Map to Enhance Students' Understanding and Promote Quality of Knowledge Structure. *The International Journal of Advanced Computer Science and Applications*, 11(4), 144–153. DOI: 10.14569/IJACSA.2020.0110420

Prasetya, D. D., Hirashima, T., Hayashi, Y. 2021. Comparing Two Extended Concept Mapping Approaches to Investigate the Distribution of Students' Achievements. *IEICE TRANSACTIONS on Information and System*, 104(2), 337–340. DOI:10.1587/transinf.2020EDL8073

Schwendimann, B. A., & Linn, M. C. 2016. Comparing two forms of concept map critique activities to facilitate knowledge integration processes in evolution education. *Journal of Research in Science Teaching*, 53(1), 70–94.

Improving Assessment and Evaluation Strategies on Online Learning – Wonorahardjo, Karmina & Habiddin (eds)

Development of virtual reality-based learning media on chemical bond materials and molecular shapes for grade 10th of senior high school students

A. Febriana, R. Joharmawan, R. Hakiki & M. Muchson*
Universitas Negeri Malang, Indonesia

ABSTRACT: The material of chemical bonds and molecular shapes is considered quite difficult for learning by students. In addition, boring and non-interactive learning media make it less interesting for students to learn. To overcome this, researchers developed VARITY (Virtual Reality Chemistry) learning media, which is an interactive learning application based on virtual reality technology with the topic of chemical bonds and molecular shapes. This study aims to conduct a media feasibility test and increase students' interest in learning. The learning media design model used is ADDIE which has five stages, namely: (1) assessment/analysis, (2) design, (3) development, (4) implementation, and (5) evaluation. Based on the results of the study, it is known that the VARITY learning media is included in the very feasible criteria in terms of material and media. The final result of the average percentage of product validity in terms of material is 87.32% (very feasible) the lack of percentage is due to very little variation in the questions in the media. Meanwhile, in terms of media, the results obtained 89.30% (very feasible), due to the lack of explanatory information related to how to use, the percentage of validity is not perfect. The media that has been revised according to the validator's suggestions and comments is tested in small groups conducted by 31 students of SMAN 1 Srengat, the results are 86.14% (very feasible). So, it can be concluded that VARITY learning media can be used in classroom learning.

Keywords: chemical bonds, learning media, molecular shapes, VARITY

1 INTRODUCTION

Chemistry is a science that has three representative levels, namely macroscopic, submicroscopic, and symbolic. The three representative levels must be considered in the chemistry learning process in the classroom so that the learning objectives can be achieved optimally (Amarlita & Sarfan 2016). Students often have difficulty in learning material that is microscopic and abstract theory. The shape of the molecule is one of the chemical materials that explains the forms of a chemical element where this material cannot be studied experimentally and the molecule itself is an invisible particle.

According to Iordache et al. (2012) one of the problems that students face in studying the topic of molecular shape is the difficulty of predicting or imagining the shape of a molecule from the point of view of three-dimensional space. Learning difficulties are a condition where students cannot learn properly, because of threats, obstacles, or disturbances in learning (Jamal 2014). Two factors cause learning difficulties, namely internal factors including students' motivation and interest in learning materials for chemical bonds and intermolecular forces, while external factors include teacher aspects, namely learning methods, teaching methods, and also the learning media used (Ahmadi & Supriyono 2013).

The lack of interest and motivation of students in learning chemistry can also be because the learning media used by the teacher is still dominated by books containing only text and pictures.

*Corresponding Author

 DOI 10.1201/9781003261346-17

The use of pictures makes students tend to be passive, less interactive because they cannot provide feedback, and less attractive to students to learn. In addition, the 2018 Global Education Census conducted by Cambridge International shows that the use of the most popular learning media in Indonesia is the blackboard, which is 92% (UCLES 2018).

One way to increase the interest and motivation of students in the learning process is to maximize learning media (Pranata et al. 2018). Learning media is a tool in the teaching and learning process and anything that can be used to stimulate the thoughts, attention, feelings, and abilities or skills of students so that it can encourage the learning process or learning activities (Shalikhah 2017). Learning media develops from time to time, along with technological developments. The development of learning media also follows the demands and needs of learning. One way to increase students' interest in the learning process is to maximize interactive learning media. Interactive learning media will certainly create a more communicative and interesting learning atmosphere between teachers and students.

In research (Merchant et al. 2013) entitled "Exploring 3-D Virtual Reality Technology for Spatial Ability and Chemistry Achievement", the results of the study concluded that students with low spatial abilities exhibited better performance while learning the use of 3D Virtual Reality compared to those with low spatial abilities students that employ 2D images. The spatial ability means to imagine something that is at the level of symbolic and submicroscopic representation or something that is abstract and more real-like (Terlecki & Newcombe 2005).

Submicroscopic and abstract chemical bonding material and molecular shapes will look interesting when presented with a 3D display in VR technology-based learning media where this media will help students predict the shape of molecules in a 3D space perspective. According to (Chou 2017) one of the benefits of using virtual reality is that it has the potential to encourage students' interest in learning.

Based on the results of the literature study, it can be seen that the use of virtual reality-based learning media on chemical bonds and molecular shapes equipped with trivia quizzes has the potential to make it easier for students to understand the material. The VR feature used in the media can help students understand the shapes of molecules in three dimensions along with related molecular information. The trivia quiz feature is expected to be used as a benchmark for the extent to which students understand chemical bonding and molecular shapes.

2 METHODS

The development model used in the development of virtual reality-based learning media on chemical bonds and molecular shapes for class X high school students uses a development model according to Lee & Owens (2004). In this development model, the media design model used is ADDIE where there are five stages, namely assessment/analysis, design, development, implementation, and evaluation.

2.1 *Assessment/analysis*

The assessment or analysis phase (assessment/analysis) aims to analyze the basic problems experienced in learning and also to determine the characteristics of the material. The researchers conducted the following steps: (1) determining the target users of the media to be developed, namely students of class X SMA, (2) determining the topics to be raised on the topic of chemical bonds and molecular shapes, (3) selecting virtual reality technology. This is the right choice to be developed in this learning media because it makes it easier for students to predict the shape of a molecule from a three-dimensional point of view.

2.2 *Design*

At the design stage, the material developed in this VR-based learning media includes chemical bonding material and molecular shapes. The storyboard of the media to be developed is made clear to facilitate the development process. In addition, the preparation of tools for validation was also carried out which included a media expert validation questionnaire and a material expert validation questionnaire which would be shown to each validator and teacher as well as a test questionnaire aimed at students where this questionnaire was used as a five-level Likert scale.

2.3 *Development*

The development of learning media is carried out as per the storyboard that has been made beforehand. The storyboard itself contains the shape of the molecule along with the information that will be displayed, the display design of the application, a list of questions and answers, and a description of the media used in the VR feature application. Next, media programming is done, molecular shape assets are created in the Blender application with .fbx format. Then proceed with compiling the appearance of the application using Unity Pro which is converted into .apk format so that it can be installed and run on Android or smartphones. The developed media was validated to determine the level of feasibility as a learning media. Validation was carried out by material experts in the field of chemistry to assess the content of the developed media, graphic design experts to assess the appearance of the media, and teachers in schools. Based on the validation results, several suggestions and comments were obtained from the validator for media improvement before the trial was carried out.

2.4 *Implementation*

The media that has been corrected based on the suggestions and comments from the validator is tested in small groups by 31 students at Senior High School 1 Srengat. The data collection instrument used was a validation questionnaire containing a checklist as well as a comment and suggestion column using a five-level Likert scale.

2.5 *Evaluation*

Researchers at this stage evaluate the learning media products. The evaluation carried out in this development is reviewing product development from the results of the trial, analyzing the test results questionnaire, and considering the comments and suggestions obtained.

3 RESULT AND DISCUSSION

Virtual Reality Chemistry (VARITY) is a virtual reality-based learning media that focuses on chemical bonds and molecular shapes. There are three-dimensional shapes of 11 molecules accompanied by brief explanations and simple practice questions that will change each time you enter the application. VARITY learning media is expected to help students understand the material

3.1 *VARITY learning media production*

The first step in making this VARITY media is to make the shape of the molecule in a three-dimensional view. Molecular shapes are created with the help of the Blender application where each molecular shape is equipped with bond angles. A blender is a pen-source software used to create three-dimensional animations (Musril et al. 2020). The final result of the molecular shape is saved in .fbx format. Next, there are 30 questions related to the topic of chemical bonds and molecular shapes with two answer choices and also brief information about each molecular shape made.

The next stage in this development is making the display of the learning media. The display of the VARITY learning media was created with the help of the Unity 3D application. Unity is a multi-platform game engine developed by Unity Technologies (Faiztyan et al. 2015). The questions and molecular shapes that have been made will then be entered and arranged one by one according to the storyboard. Giving background music, giving icons, and placing text are also done with the help of this application.

3.2 *Description of VARITY learning media*

The final result of developing VARITY learning media in the form of an application with a size of 94 Mb can be run on a smartphone/android with a minimum specification of android version 7.0 (Nougat). The application can be downloaded by the google drive link and installed independently. Where once installed the application does not require an internet connection to run the application. It is recommended to run VARITY to use VR Box to get a new experience that is more exciting and interesting.

The initial display of VARITY has 4 options, namely molecular info, quiz, settings, and more info as well as an exit button to exit the application. The way to select the menu is by pointing the red dot towards the desired menu and waiting for approximately 2 seconds. In the molecular info menu, there is a three-dimensional display of 11 molecular shapes that are placed throughout the room where the room is made to resemble a classroom with a blackboard on the front. According to Roffiq et al. (2017) the atmosphere of the learning environment affects learning outcomes. In the quiz menu section, there are 5 questions with two answer choices that students can do to test the extent of students' knowledge regarding the topic of chemical bonds and molecular shapes. The questions that appear will be randomized automatically when you enter the application, so that every time you enter the application, the questions you get will be different. In the settings menu, there are music buttons and audio buttons. If students feel disturbed by background music or the sound of the red button, the sound can be turned off in the settings menu. The last menu on the VARITY learning media covers more information. In this menu there are two options, namely a description which contains a brief description of VARITY and how to use it, then the about option contains identity.

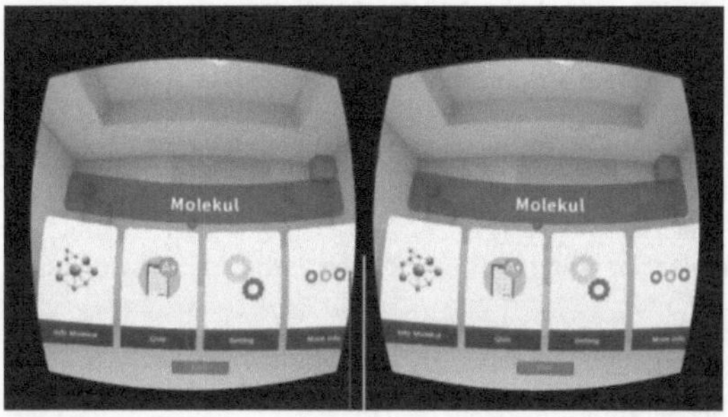

Figure 1. The initial display of VARITY.

Figure 2. Molecular info menu display.

Figure 3. Display menu quiz.

3.3 *VARITY feasibility in terms of learning media*

Media VARITY (Virtual Reality Chemistry) in terms of media can be said to be very feasible to use in learning activities. Based on the results of filling out a questionnaire conducted by media experts and two high school teachers, the average percentage of product validation results as a learning medium was 89.30%. When compared with the eligibility criteria in Table 1, it can be concluded that the VARITY learning media is very feasible to use and very valid.

Table 1. Results of validation of development products as learning media.

No.	Criteria Assessed	Percentage (%)	Criteria
1.	Overall Display of Physical Media	89.99%	Very feasible
2.	VARITY Application Menu Display	88.62%	Very feasible
Average Percentage		89.30%	**Very feasible**

3.4 *VARITY feasibility in terms of material*

VARITY learning media validation was also carried out in terms of material. Validation was carried out by material experts, namely a Lecturer of Inorganic Chemistry, State University of Malang, and two high school teachers. The results obtained are as follows.

Table 2. Results of validation of development products as learning media.

No.	Criteria Assessed	Percentage (%)	Criteria
1.	The suitability of the questions on the quiz with the material	86.66%	Very feasible
2.	Ease of understanding the questions on the quiz	80.00%	Feasible
3.	Suitability of molecules on the application form with the material	86.66%	Very feasible
4.	Suitability of the bond angles in applications with materials	86.66%	Very feasible
5.	Suitability boiling point molecules in applications with materials	86.66%	Very feasible
6.	Compatibility of the melting point of the molecule on application to the material	86.66%	Very feasible
7.	Compatibility of molecular polarity on application with material	86.66%	Very feasible
8.	The suitability of the molecular coordination number in the application with the material	86.66%	Very feasible
9.	The truth of the quiz questions in testing the concept	93.33%	Very feasible
10.	The truth of the answers to the questions on the quiz	93.33%	Very feasible
Average Percentage		87.32%	**Very feasible**

Based on the results of Table 2, the average percentage of learning media validation in terms of material is 87.32%. When compared with the eligibility criteria in Table 1, it can be concluded that the VARITY learning media is very valid and very feasible to use to improve students' understanding of the topic of molecular shapes.

3.5 *Results of small group trials*

The small group trial was carried out at SMA Negeri 1 Srengat by 31 students of class X MIPA. Before the trial questionnaire was distributed, students were shown a short video explaining how to use the VARITY learning media. Then students were asked to download VARITY from the Google Drive link that was provided and test questionnaires were distributed after students tried the application.

Figure 2. Students try to run VARITY media.

Table 3. Results of small group trials.

No	Criteria Assessed	Percentage (%)	Criteria
1.	Media Presentation	86.06 %	Very feasible
2.	Information Clarity	84.35%	Very feasible
3.	Program Usage	85.15%	Feasible
4.	Media Effectiveness (according to students)	89.02%	Very feasible
Average Percentage		86.14%	**Very feasible**

The percentage of media effectiveness criteria obtained the greatest results, 89.02%. This shows that the VARITY media for students can increase their effectiveness and desire in learning. As for the final result data, the average percentage obtained from the small group trial shown in Table 3 is 86.14%. When compared with the eligibility criteria of Table 1, it can be said that the VARITY learning media is valid and suitable for students to use in learning. These results indicate that students feel that the existence of VARITY media can increase students' interest in learning and make it easier for students to understand the material.

4 CONCLUSION

Learning media VARITY (Virtual Reality Chemistry) is a learning media based on virtual reality technology on the topic of chemical bonds and molecular shapes. The minimum android/smartphone specification that can be used to run this learning media is Android version 7.0 (Nougat). In addition, the use of VARITY is also equipped with a VR Box to get a new experience in learning. Eleven molecular shapes are displayed in three-dimensional form accompanied by brief information related to the molecule. Equipped with a trivia quiz that can be done by students where the five questions that appear will continue to change every time they enter the application. The average percentage of product validity results in terms of material is 87.32% (very feasible), in terms of media 89.30% (very feasible) and from the results of small group trials of 86.14% (very feasible) so that it can be it is said that the VARITY learning media is very feasible to be applied in learning. These results indicate that according to students' perceptions, the existence of VARITY learning media can increase students' interest in learning and motivation in understanding the topic of chemical bonds and molecular shapes.

REFERENCES

Ahmadi, A., & Supriyono, W. (2013). PsikologiBelajar. Rineka Cipta.

Amarlita, D. M., & Sarfan, E. (2016). Analisis Kemampuan Makroskopis, Mikroskopis Dan Simbolik Pada Materi Kesetimbangan Kimia. *BIMAFIKA Jurnal MIPA, Kependidikan Dan Terapan*, *0271*, 1–19. http://pbs.fkip.unila.ac.id/wp-content/uploads/sites/90/2015/03/Blangko_Ujian_Skripsi.doc

Chou, C. C. (2017). An analysis of the 3D video and interactive response approach effects on the science remedial teaching for fourth grade underachieving students. *Eurasia Journal of Mathematics, Science and Technology Education*, *13*(4), 1059–1073. https://doi.org/10.12973/eurasia.2017.00658a

Faiztyan, I. F., Isnanto, R. R., & Widianto, E. D. (2015). Perancangan dan Pembuatan Aplikasi Visualisasi 3D Interaktif Masjid Agung Jawa Tengah Menggunakan Unity3D. *Jurnal Teknologi Dan Sistem Komputer*, *3*(2), 207. https://doi.org/10.14710/jtsiskom.3.2.2015.207-212

Iordache Dr, D. D., Pribeanu, C., & Balog, A. (2012). Influence of specific ar capabilities on the learning effectiveness and efficiency. *Studies in Informatics and Control*, *21*(3), 233–240. https://doi.org/10.24846/v21i3y201201

Jamal, F. (2014). Analisis Kesulitan Belajar Siswa Dalam Mata Pelajaran Matematika Pada Materi Peluang Kelas XI IPA SMA Muhammadiyah Meulaboh Johan Pahlawan. *Jurnal MAJU (Jurnal Pendidikan Matematika)*, *1*(1), 18–36. http://www.ejournal.stkipbbm.ac.id/index.php/mtk/article/view/232

Lee, W. W., & Owens, D. . (2004). *Multimedia-Based Instructional Design* (Second Edi). Pfeiffer.

Merchant, Z., Goetz, E. T., Keeney-Kennicutt, W., Cifuentes, L., Kwok, O., & Davis, T. J. (2013). Exploring 3-D virtual reality technology for spatial ability and chemistry achievement. *Journal of Computer Assisted Learning*, *29*(6), 579–590. https://doi.org/10.1111/jcal.12018

Musril, H. A., Jasmienti, J., & Hurrahman, M. (2020). Implementasi Teknologi Virtual Reality Pada Media Pembelajaran Perakitan Komputer. *Jurnal Nasional Pendidikan Teknik Informatika (JANAPATI)*, *9*(1), 83. https://doi.org/10.23887/janapati.v9i1.23215

Pranata, M. A., Santyadiputra, G. S., & Sindu, I. G. P. (2018). Rancangan Game Balinese Fruit Shooter Berbasis Virtual Reality Sebagai Media Pembelajaran. *Jurnal Nasional Pendidikan Teknik Informatika (JANAPATI)*, *6*(3), 256. https://doi.org/10.23887/janapati.v6i3.11994

Roffiq, A., Qiram, I., & Rubiono, G. (2017). Media Musik Dan Lagu Pada Proses Pembelajaran. *JPDI (Jurnal Pendidikan Dasar Indonesia)*, *2*(2), 35. https://doi.org/10.26737/jpdi.v2i2.330

Shalikhah, N. D. (2017). Media Pembelajaran Interaktif Lectora Inspire sebagai Inovasi Pembelajaran. *Warta LPM*, *20*(1), 9–16. https://doi.org/10.23917/warta.v19i3.2842

Terlecki, M. S., & Newcombe, N. S. (2005). How important is the digital divide? the relation of computer and videogame usage to gender differences in mental rotation ability. *Sex Roles*, *53*(5–6), 433–441. https://doi.org/10.1007/s11199-005-6765-0

UCLES. (2018). Global education census report. *Global Education Census Report 2018.*, 48. https://www.cambridgeinternational.org/Images/514611-global-education-census-survey-report.pdf

Improving Assessment and Evaluation Strategies on Online Learning – Wonorahardjo, Karmina & Habiddin (eds)

The Kirkpatrick model integration as an evaluation design for the implementation of vocational educator training in the 21st century

E. Sutadji*, D.E. Rahayu, N. Fadilah, T. Sukmasari, A.B.N.R. Putra & A.M. Rini
Universitas Negeri Malang, Indonesia

ABSTRACT: Education and training can be interpreted as acquiring knowledge, skills, and attitudes that enable humans to achieve individual and organizational goals today and in the future. Meanwhile, work productivity is lost when employees attend training. To ensure that the training program provides benefits, it is necessary to evaluate the training program. The objectives of this study include: (1) to measure the satisfaction of vocational educators at the reaction level; (2) to measure the increase in knowledge and skills of vocational educators at the learning level; (3) to measure the ability of vocational educators in applying knowledge at the behavioral level; and (4) to measure changes in the ability of vocational educators at the impact level. The research method used in this study is an evaluation research method. Evaluation research methods include: (1) component identification; (2) identification of aspects; (3) identification of indicators; (4) determination of the data source; (5) determination of the data collection method; (6) determination of the data collection instrument. The evaluation model used is the Kirkpatrick model, consisting of input, process, output, and outcomes. The instruments used in this study were questionnaires, learning outcome tests, interviews, and documentation studies. Data analysis was done using triangulation method. The results of this study include: (1) the average percentage level of reaction to the resource component is 82%, the training material component is 87.7%, the facility component is 85.7%, the activity schedule component is 91.57%; (2) increased knowledge and skills of vocational educators at the learning level by 79%; (3) the ability of vocational educators to apply knowledge at the behavioral level is in the excellent category; and (4) at the impact level, positive changes occurred after the training, as evidenced by the fact that vocational educators were able to prepare annual programs, prepare semester programs, prepare lesson plans, and prepare worksheets and get an excellent category.

Keywords: Kirkpatrick model, evaluation design, vocational educator, 21st century

1 INTRODUCTION

Training is one of the efforts in improving the quality of human resources. Developed countries give great importance to the quality of human resources, and hence education and training is a priority program (Pilz & Regel 2021; Wu 2021). In many ways, education and training are considered human capital that will improve the organization's quality. On the other side, many human resource experts consider the need to evaluate training as part of quality control in the quality assurance process. One form of quality control in training activities is to measure the success rate of education and training. Control on the implementation of the training is intended to determine the achievement of the programs that have been held. Education and training can be interpreted as acquiring knowledge,

*Corresponding Author

DOI 10.1201/9781003261346-18

skills, and attitudes that enable humans to achieve individual and organizational goals today and in the future (Khlaisang & Songkram 2019; Kim et al. 2019).

The Law on State Civil Apparatus states that the State Civil Apparatus as a profession is based on principles, one of which is having the necessary competencies following the field of duty. By Article 70 of the Law on State Civil Apparatus, as an effort to develop competence for the ASN, one of the ways to do this is through education and training activities. Human resource training and development are defined as processes that utilize various methods to provide the skills needed for new and existing employees in carrying out their work (Basaran et al. 2021; Haq et al. 2019). This definition is in line with the definition of training and education in the 2010 Government Regulation on Education and Training of Civil Servants, namely the process of organizing teaching and learning to improve civil servants' ability.

Training for employees has several functions. The training aims to develop employees' knowledge of the company's culture and its competitors and assist skilled workers in working with new technologies. In addition, to assist employees in understanding how to work efficiently and effectively in a team that aims for quality products and services and ensures a corporate culture that emphasizes innovation, creativity, and knowledge (León et al. 2021; Vilà-Giménez & Prieto 2021). Meanwhile, work productivity is lost when employees attend training. To ensure that the training program provides benefits and is not in vain, it is necessary to evaluate the training program. An evaluation of the impact and effectiveness of the training is needed so that the strengths and weaknesses in the program can be identified for improvements. This is following the proposed evaluation function, namely to obtain accurate and objective information on a program that has been planned and implemented in the previous phase (Peñalba & Clacio 2017; Haq et al. 2019).

In line with the definitions and evaluation functions of the various opinions above, in (Law of 2003) article number 20 concerning the National Education System Article 57 paragraph (1), it is stated that evaluation is carried out in the context of controlling the quality of education nationally as a form of accountability of education providers to interested parties, including participants, educational institutions, and programs. Program evaluation models can be grouped, namely (1) Goal-Oriented Evaluation, (2) Decision-Oriented Evaluation, (3) Transactional Evaluation, (4) Goal-Free Evaluation, (5) Adversary Evaluation, and (6) Kirkpatrick Model Evaluation (Alsalamah & Callinan 2021; Terttiaavini et al. 2020). According to the Law on Teachers and Lecturers Number 14 of 2005, Article 10 states that pedagogic competence can manage student learning. These competencies include understanding students, designing and implementing learning, evaluating learning outcomes, and developing students to actualize their various potentials (Kunandar 2007). Efforts to increase teachers' ability in making learning tools require coaching through socialization and training, both from school supervisors and from school principals. One of the forms is by holding a Training for Preparation of Learning Devices for making learning devices (Singh et al. 2021; Wiratraman & Lafrance 2021; Zulkardi et al. 2021). The Learning Device Preparation Training was chosen because, in this training method, theory and practice are balanced. In other words, participants directly apply the theory presented in the form of the expected product. In addition, its implementation emphasizes the implementation of the learning process well. Socialization and training, such as the Training for Preparation of Learning Devices, need to be directed at efforts that are providing opportunities for teachers so that the potential of teacher resources can grow and develop so that they are better able to carry out their primary task, namely carrying out a quality learning process (Anis & Anwar 2020; Masitoh & Cahyani 2020).

Based on the results of observations of the Main Learning Plan (RPP) document at SMK Negeri 1 Boyolangu, it was found that the components in the RPP and LKS differ from one teacher to another. This shows that the training results have not been able to describe the urgency of the training. Information about the impact of training on improving teacher performance in compiling learning tools has also not been found. Training as a system must show a systematic evaluation model of training programs in which the outputs and outcomes show the competence of their respective substances. So far, the evaluation carried out on the training held at SMK Negeri 1 Boyolangu still does not have a measuring tool that can show the success of the training and the impact of the training. For this reason, it is necessary to apply one of the training evaluation models, namely the Kirkpatrick evaluation model, to measure the success of the Training for Learning Device Preparation to know the impact of the training that has been followed by the teachers on the ability to develop learning tools.

2 METHOD

Research Methods: The research method used in this study is an evaluation research method. Transformation means changing shape. The learning process means an effort to change students who have initially been still in a state of not knowing the knowledge given by the teacher. After going through the learning process, it is expected to know well. The training in the preparation of learning tools referred to in this study serves to improve teacher competence in the field of preparing learning materials in the form of semester programs, annual programs, lesson plans, and student worksheets.

Research methods are needed to achieve research objectives. This research was conducted to explore data intensively and carefully analyze the effectiveness of training in the preparation of learning tools at SMK Negeri 1 Boyolangu Tulungagung. The evaluation model used is the Kirkpatrick model, consisting of input, process, output, and outcomes. The research design of training effectiveness using Kirkpatrick evaluation aims to reveal all components related to training, which consist of reaction (reaction), learning (learning outcomes), behavior (behavior), and result (impact).

The subjects of this study were teachers of SMK Negeri 1 Boyolangu who participated in the training activities. Respondents were 48 teachers consisting of 5 vice principals, 10 teachers in groups A and B (other than productive teachers), 33 teachers representing 11 competency skills. Kirkpatrick's evaluation consists of 4 levels, namely reaction, learning, behavior, and result. Each of these stages must be carried out sequentially because each level affects the next stage. The instruments used in this study were questionnaires, learning outcome tests, interviews, and documentation studies. The credibility test in this study uses triangulation, namely, to test the credibility of the data by checking the data on the same data source with different techniques. The analysis in this study uses the step of calculating the total score from the data that has been matched between the questionnaire data with documentation and interviews, then determining the percentage of implementation of each sub-indicator.

3 RESULTS

The training evaluation was conducted on 48 respondents consisting of 9 teachers in groups A and B (other than productive teachers); 39 teachers represent 11 competency skills in SMK Negeri 1 Boyolangu Tulungagung. The description of Kirkpatrick's evaluation results for each level can be explained as follows.

3.1 *Reaction aspect*

The reaction measures participants' satisfaction about activities during training (training) Preparation of Learning Devices at SMK Negeri 1 Boyolangu. The evaluation results can be seen through a closed questionnaire instrument from the answers given by respondents through Google Forms. Information can be provided by participants about resource persons, materials, facilities, and schedule of activities. The organizers hope that the information received can improve the effectiveness of training and services in the next activity. Each aspect of the reaction are described in the following categories.

Table 1. Reaction success criteria.

No	Percentage	Category
1.	91–100%	Very good
2.	81–90%	Well
3.	71–80%	Currently
4.	61–70%	Not enough
5.	≤60%	Very less

The results of processing the data description of the effectiveness of the Preparation of Learning Devices training at SMK Negeri 1 Boyolangu on the resource components are as follows. The resource persons with the appropriateness of expertise component showed that the participants' reactions to the resource persons were in the moderate category 0%, the good category 85.4%, and the very good category 14.6%. The resource persons with class mastery component showed that the participants' reactions to the resource persons were in the moderate category 10.4%, the good category 79.2%, and the very good category 10.4%. Resource persons with skills component involving participants showed that participants' reactions to resource persons were in the medium category 8.3%, in the good category 83.3%, and the very good category by 8.3%.

The component of the suitability of the material with the training objectives showed that the participants' reactions to the suitability of the training material provided were in the less than 0% category, 91.7% in the good category, and 8.3% in the excellent category. The material's suitability with the training participants' needs showed that the participants' reaction to the participants' needs for training materials was in the poor category 2.1%, good category 95.8%, and excellent category 2.1%. The reaction of the training participants to the component of mastery of the material shows that the participants' reaction to the mastery of the training material is in the poor category 16.7%, good category 75%, and excellent category 8.3%. The facility component is the opinion of the training participants relating to the facilities and infrastructure provided by the organizers for the training participants, namely the hall, focus, sound system, chairs, air conditioning, soft copy of training materials, prayer room, toilet, morning snack, and lunch. The participants' reactions showed that participants who stated less were 8.3%, good 85.81%, and excellent 8.7%. In the activity schedule component, the aspect to be evaluated is the opinion of the training participants regarding the suitability of starting training (training) and ending activities for each session. The results of the participants' reactions stated that those who chose the poor category were 9.73%, 91.57% right, and excellent 2.1%.

3.2 *Learning aspect*

At the learning stage (learning outcomes), the effectiveness of teacher training at SMK Negeri 1 Boyolangu Tulungagung was measured using the test results at the end of the activity. Competencies possessed by participants are calculated by the percentage of scores obtained by participants through the test, as has been written in the theory that the test results of participants are given a score and classified according to the Badrujaman (2016) category.

Table 2. Learning categories.

No	Percentage	Category
1.	75–100	Well
2.	51–75	Moderate/enough
3.	25–50	Low
4.	0–25	Very low

Based on the data tabulation above, training participants can be classified as follows.

Table 3. Classification of test result categories.

No.	Category	Amount	Percentage
1.	Good	38 people	79%
2.	Moderate/enough	4 people	8%
3.	Low	6 people	13%
4.	Very low	0 people	0%

The test results from the most training participants were in a suitable category, namely 79%. Furthermore, it will also be described based on four components of training material, namely, preparation of annual programs, preparation of semester programs, preparation of lesson plans, and preparation of worksheets.

3.3 *Behavior aspect*

Based on the analysis of the questionnaire given to the alumni of the Learning Device Preparation training program, it was found that the general application of the results of interviews was about events, activities, organizations, feelings, motivations, demands, and concerns for the school. According to the findings and data analysis, it is known that the indicators from the assessment of the training participants through the perceptions of superiors and the assessment of the training participants through the perceptions of colleagues indicate the qualifications of "Very Good", which occupies the highest score, namely the indicators that after attending the training participants are able to compile annual programs, develop semester program, compile RPP, and compile LKS. Furthermore, the results are shown in the following table.

Table 4. Behavioral success criteria.

No.	Percentage	Category
1	91–100%	Very good
2	81–90%	Well
3	71–80%	Moderate/Enough
4	61–70%	Not enough
5	$\leq$ 60%	Very less

3.4 *Result aspect*

Based on the documentation results that the researchers completed on April 12, 2021, at 09.00–3.00 at SMK Negeri 1 Boyolangu Tulungagung. The results of the researcher's documentation comprised data from the curriculum and the data attached to the back. The results of the percentage data for each aspect studied in the Training Evaluation Using the Kirkpatrick Model in Preparation of Learning Devices at SMK Negeri 1 Boyolangu Tulungagung were then described in the form of a bar chart, which is presented briefly in the table.

Table 5. Documentation before training.

No.	Rated aspect	Present (%)	Category
1	Vision and mission	79	Currently
2	Technical Instructions Document	79	Currently
3	Participant Biodata	73	Currently
4	List of Attendees	79	Currently
5	Training Material Archive	73	Currently
6	Evaluation Result Archive	73	Currently

4 DISCUSSION

The evaluation research of Learning Device Preparation Training at SMK Negeri 1 Boyolangu is an evaluative research model using the Kirkpatrick evaluation model. This model requires researchers to research systematically, starting from the reaction, learning, behavior, and impact stages (Cahapay 2021; Sahni 2020). This is done because the results of the evaluation of each stage will affect the results of the evaluation at the next stage.

The discussion of the components of the reactions or responses of the training participants in the Learning Device Preparation Training at SMK Negeri 1 Boyolagua focused on the sources, materials, facilities, and training schedule. Based on the data above, it can be concluded that the average reaction (reaction) of the training participants to the resource person is 82.63% (good), reaction (reaction) to the material is 87.5% (good), reaction (reaction) to the material is 87.5% (good) to the facility 85.83% (good), and reaction (reaction) to the schedule 91.66% (very good). The achievement of the reaction component of the training participants to the resource persons (82.63%) was classified in the better category compared to Juanda (2011) because the results of his research were 43% of participants who stated that the resource persons were good, this was due to the lack of readiness and quality, namely one was able to prepare 3–4 training materials. Likewise, Herawaty's research (2012) stated that 65% of participants said they were good because there were several complaints from participants, namely some of the speakers were less communicative, had boring presentations, and lacked expertise in some material. Meanwhile, the research on Preparation of Learning Devices at SMK Negeri 1 Boyolangu compared to Kurnia's research (2015) was also better because his research on resource persons was on average (55%).

The achievement of the components of the reaction of the trainees to the training material (87.5%) was classified in the better category compared to Juanda (2011), who stated less (60%) because the material provided was less relevant to the needs of the training participants. Administration while the majority of participants are educators. The results of Herawaty's research (2012), 53% said it was good because the training materials were in accordance with the needs of participants in the world of work. Meanwhile, 54% of Kurnia's research stated that it was moderate because a lot of the material provided could not guarantee that participants could master the material taught during industrial work practices. The achievement of the reaction component of the training participants to the training facilities (85.83%) is classified in the better category by researcher Juanda (2011). Similarly, compared to research by Herawaty (2012), 73% is classified as good because of the availability of adequate classrooms, room temperature, and the practice tools used are very standard. In contrast to the results of Kurnia's research (2015), which states that it is very lacking (35%), this is indicated by poor internet access, cleanliness, state of toilets, and places of worship that need attention. The achievement of the components of the reaction of the trainees to the training schedule (91.6%) was classified in the better category with the previous researcher, namely Juanda (2011), who stated that 65% were in a good category because the addition of time, which was initially designed 18 days to 24 days was by the material training. Likewise, as per Herawaty (2012), 77% stated that the training time was good for participants in choosing the time for training and the schedule of training events.

In the discussion of the learning stage, it was carried out to find out the learning outcomes, which include increasing the trainees' knowledge and skills on the material delivered by the resource persons. This is measured by using the test results at the end of the activity (Bari et al. 2021; Reese et al. 2021). This is in line with Kirkpatrick's statement, which states that learning measurement is essential because there will be no change in behavior (level 3) without the achievement of one of the objectives of the learning program either from knowledge, skills, or attitudes (Lantu et al. 2020; Terttiaavini et al. 2020). Participants' learning outcomes can be seen from changes in participant behavior after returning to school in applying the training results to students. Based on the data above, it can be concluded that the participants in the training learning (learning) are in good category 79%, moderate/enough category 8%, low category 13%, and deficient category 0%.

The evaluation of the Learning Device Preparation Training at SMKN 1 Boyolangu showed a change in the ability to apply knowledge in schools; either in knowledge, skills, or attitudes. The application of this knowledge indicates a positive behavior change in the participants. According to the findings and data analysis, it is known that the indicators from the assessment of the training participants through the perception of their superiors show the qualifications of "Very Good", which occupies the highest score, namely the indicator that after attending the training participants become able to compile annual programs, arrange semester programs, prepare lesson plans, and compile worksheets. In addition, participants in the training evaluation experienced a change in behavior/attitude. Changes in behavior in question are changes in positive behavior in events, activities, organizations, feelings, motivations, demands, and relatively permanent concerns to impact students positively (Mohammed Benmoussa 2020; Sahni 2020). This level evaluation cannot be carried out before the first and second-level evaluations are carried out. Even when the satisfaction index is excellent (level 1) and learning objectives are met or achieved (level 2), knowledge transfer into behavior may not take place.

The discussion of the result component or documentation of the training participants in the Learning Device Preparation Training at SMK Negeri 1 Boyolagua focused on the Vision and Mission, Technical Instruction Documents, Participant Biodata, Attendance List, Research Material Archives, and Evaluation Results Archives. Based on the data above, it can be concluded that the Vision and Mission documentation is 85% (Good), Technical Guidance Documents is 85% (Good), Participants Biodata is 85% (Good), Attendance List is 83% (Good), Research Material Archive is 83% (Good), and Evaluation Result Archive is 83% (Good). Based on the documentation results that the researchers completed on April 12, 2021, at 09.00–3.00 at SMK Negeri 1 Boyolangu Tulungagung. The results of the researcher's documentation took data from the curriculum and the data attached to the back.

The results of the percentage data for each aspect studied in the Training Evaluation Using the Kirkpatrick Model in Preparation of Learning Devices at SMK Negeri 1 Boyolangu Tulungagung are then described in a bar chart; the data was presented briefly in the table above.

The achievement of the result component (Impact) of training on post-training is classified in a good category with the previous researcher, namely Nurhayati (2018), who stated that the post-training curriculum (training schedule, training materials, benefits of training, and extracurricular materials) was 89.55 (Good category). Likewise, Sandra (2019) 81.09% stated that participants had written indicators well, it would be easy to write down their goals because writing indicators still needed to be improved, so writing goals still needed more intensive practice.

5 CONCLUSION

This research resulted in several conclusions. First, the average percentage level at the reaction level at the resource component is 82%, the training material component is 87.7%, the facility component is 85.7%, the activity schedule component is 91.57%. Second, the increase in knowledge and skills of vocational educators at the learning level by 79%. Third, the ability of vocational educators to apply knowledge at the behavioral level is in the excellent category. Fourth, there are positive changes after the training at the impact level, as evidenced by the fact that vocational educators can prepare annual programs, prepare semester programs, prepare lesson plans, prepare worksheets, and get an excellent category.

REFERENCES

Alsalamah, A. and Callinan, C. (2021) 'Adaptation of Kirkpatrick's four-level model of training criteria to evaluate training programmes for head teachers', *Education Sciences*, 11(3).

Anis, M. and Anwar, C. (2020) 'Self-organized learning environment teaching strategy for ELT in Merdeka Belajar concept for high school students in Indonesia', *JEES (Journal of English Educators Society)*, 5(2), pp. 199–204.

Bari, S. *et al.* (2021) 'Surgical data strengthening in Ethiopia: results of a Kirkpatrick framework evaluation of a data quality intervention', *Global Health Action*, 14(1).

Basaran, M. *et al.* (2021) 'Evaluation of Preschool Education Program According to CIPP Model', *Pedagogical Research*, 6(2), p. em0091.

CAHAPAY, M. (2021) 'Kirkpatrick Model: Its Limitations as Used in Higher Education Evaluation', *International Journal of Assessment Tools in Education*, 8(1), pp. 135–144.

Haq, A., Slamet, A. and Suminar, T. (2019) 'Link and Match Based Partnership Management in the Fashion Expertise Program at SMK Negeri 6 Semarang', *Educational Management*, 8(2), pp. 194–200.

Khlaisang, J. and Songkram, N. (2019) 'Designing a Virtual Learning Environment System for Teaching Twenty-First Century Skills to Higher Education Students in ASEAN', *Technology, Knowledge and Learning*, 24(1), pp. 41–63.

Kim, S., Raza, M. and Seidman, E. (2019) 'Improving 21st-century teaching skills?: The key to effective 21st-century learners', *Research in Comparative & International Education*, 14(1), pp. 99 –117.

Lantu, D. C. *et al.* (2020) 'Was the training effective? Evaluation of managers' behavior after a leader development program in Indonesia's best corporate university', *International Journal of Training Research*, pp. 1–16.

León, S. P., Carcelén Fraile, M. del C. and García-Martínez, I. (2021) 'Development of Cognitive Abilities through the Abacus in Primary Education Students: A Randomized Controlled Clinical Trial', *Education Sciences*, 11(2), p. 83.

Masitoh, S. and Cahyani, F. (2020) 'Penerapan Sistem Among Dalam Proses Pendidikan Suatu Upaya Mengembangkan Kompetensi Guru', *Kwangsan: Jurnal Teknologi Pendidikan*, 8(1), p. 122.

Mohammed BENMOUSSA (2020) 'Evaluation Of Employees Training Effectiveness In Algerian Universities Using Kirkpatrick Model: An Exploratory Study', *Journal of Economics and Applied Statistics*, 17(1), pp. 6–23.

Peñalba, E. H. and Clacio, D. M. (2017) 'Assessment of Special Provision for Empowering Diverse Pupils (SPED) Extension Service Project using Context, Input, Process, and Product Evaluation Model (CIPP) Ericson H. Peñalba & Denia M. Clacio', *Faculty Journal*, 6(1), pp. 15–30.

Pilz, M. and Regel, J. (2021) 'Vocational Education and Training in India: Prospects and Challenges from an Outside Perspective', *Margin*, 15(1), pp. 101–121.

Reese, D. *et al.* (2021) 'Quality improvement education innovation: evaluation of Coursera MOOC "Take the Lead on Healthcare Quality Improvement"', *Journal of Research in Nursing*, 26(1–2), pp. 62–78.

Sahni, J. (2020) 'Managerial training effectiveness: An assessment through Kirkpatrick framework', *TEM Journal*, 9(3), pp. 1227–1233.

Singh, S., Sharma, L. and Kumar, B. (2021) 'Machine Learning Based Predictive Model for Coronavirus Pandemic Machine Learning Based Predictive Model for Coronavirus Pandemic', *Journal of Physics: Conference Series PAPER*.

Terttiaavini *et al.* (2020) 'Evaluating the Kemplang Tunu Production Training for Low Income and Education Communities Using the Kirkpatrick Model', *Advances in Economics, Business and Management Research*, 151(1), pp. 238–242.

Vilà-Giménez, I. and Prieto, P. (2021) 'The Value of Non-Referential Gestures: A Systematic Review of Their Cognitive and Linguistic Effects in Children's Language Development', *Children*, 8(2), p. 148.

Wiratraman, H. P. and Lafrance, S. (2021) 'Protecting Freedom of Expression in Multicultural Societies: Comparing Constitutionalism in Indonesia and Canada', *Yuridika*, 36(1), p. 75.

Wu, X. (2021) 'Application of Artificial Intelligence in Modern Vocational Education Technology', *Journal of Physics: Conference Series*, 1881(3).

Zulkardi, Nusantara, D. S. and Putri, R. I. I. (2021) 'Designing PISA-like task on uncertainty and data using Covid-19 context', *Journal of Physics: Conference Series*, 1722(5), pp. 1–6.

Improving Assessment and Evaluation Strategies on Online Learning – Wonorahardjo, Karmina & Habiddin (eds)

Identification of intermolecular force misconceptions in students with different scientific thinking skills and the improvement efforts using conceptual change text

D.M. Ma'rufah*, Effendy & S. Wonorahardjo
Universitas Negeri Malang, Indonesia

ABSTRACT: Intermolecular forces are part of chemistry lessons at senior high school and Madrasah Aliyah (MA), which include many abstract concepts and macroscopic, microscopic, and symbolic representations that cause difficulties for students in learning. These difficulties can lead to misunderstandings that lead to misconceptions if they occur consistently. This study aims to: (1) find out the level of development of the student scientific thinking skills (STS); (2) identifying students' misconceptions about intermolecular force; (3) knowing the ease of eliminating misconceptions in students with different scientific thinking skills (STS) and (4) knowing the effectiveness of the conceptual change text in eliminating intermolecular force misconceptions. The method used a one-group pretest-posttest design model with descriptive analysis. The results of the study are as follows. First, 89.1% of grade XI high school students were at the STS concrete level, and 10.9% of students are at low formal level. Second, the misconceptions identified were: (a) there was a covalent bond break between the O and H atoms in the water molecule due to heating; (b) intermolecular forces are stronger than intramolecular forces; (c) the surface shape of concave fluids is influenced by dipole–dipole intermolecular force; (d) in the smelting and freezing process, the volume of ice and water is the same; and (e) substances that have strong intermolecular forces have large vapor pressures. Third, students with higher STS levels tend to more easily eliminate misconceptions. Fourth, the conceptual change text can be used to address the misconceptions that students have.

Keywords: misconception, scientific thinking skill, conceptual change text, intermolecular force

1 INTRODUCTION

The topic of intermolecular forces is a basic and essential concept in understanding other chemical concepts, one of which is the colligative properties of solutions. Moreover, intermolecular forces are used to understand the physical and chemical properties of substances, such as melting point, boiling point, fluidity, and acid–base properties (Nicoll 2001; Tarhan, et al. 2008). The concept of intermolecular forces is abstract, far from students' daily experiences, they may have difficulty in understanding the concepts, and as a result, alternative conceptions are easy to understand (Coll & Taylor 2002; Gabel 1999; Rompayom et al. 2011; Taber & Coll 2002; Tan & Treagust, 1999; Unal et al. 2006). Misconceptions are concepts that are different from the generally accepted scientific understanding (Ozmen 2004) and occur consistently.

To understand abstract concepts, it is necessary to have the ability to understand concepts at the macroscopic, sub-microscopic, and microscopic levels. To master the concept of intermolecular forces from concrete concepts to abstract concepts, formal thinking skills are required. The pattern of thinking in formal operations is a constituent of scientific thinking skills. Scientific thinking

*Corresponding Author

DOI 10.1201/9781003261346-19

skills are generally owned by students who have reached the level of formal thinking that begins to develop and is found in children at the age of 12–15 years (Piaget, 1958), or equivalent to junior high school students. However, in reality not all students in that age group develop formal thinking.

The causes of misconceptions in students are limitations in building understanding of a concept they receive during the learning process and inability to describe chemistry at the macroscopic, sub-microscopic, and symbolic levels (Barke et al. 2009; Chittleborough 2009). Several studies show misconceptions about the nature and differences of intermolecular and intramolecular forces, the breakdown of covalent bonds due to heating, differences in the boiling point of compounds, and their relationship with intermolecular and intramolecular forces (Affifanur 2012; Rompayom et al. 2011; Tan & Treagust 1999). Misconceptions must be eliminated so that they are not firmly embedded and integrated in a student's cognitive structure, which can interfere with further learning (Treagust 2006). Misconceptions in principle can be overcome with a concept change strategy. Concept change text is designed to encourage students change their prejudices and is based on Piaget's ideas of assimilation, accommodation, and imbalance (Wang & Andre 1991). This concept is designed to change students' misconceptions by encouraging students to express their conceptions, generating dissatisfaction, and followed by correct explanations that are easy to understand and make sense for students. This concept change text is effective in creating concept change and leads to meaningful learning of science concepts (Yilmaz et al. 2001).

2 RESEARCH METHODS

2.1 *Research design*

This study used a one-group pretest-posttest design with descriptive analysis. Descriptive design is used to describe identification of the level of development of Scientific Thinking Skills (STS), identification of students' misconceptions about intermolecular forces, and the ease of elimination of misconceptions in students with different STS.

2.2 *Research instruments*

In this study, there are two types of research data that will be collected, STS data and students' misconceptions of intermolecular forces. Data was collected by using the STS test and a two-tier multiple-choice test on intermolecular forces.

2.3 *Student STS test*

In this study, to measure students' STS, the revised 2000 edition of the Classroom Test of Scientific Reasoning (CTSR) instrument was used, which was developed by Lawson. The STS of students is determined based on the criteria shown in Table 1.

Table 1. Criteria for student STS level based on *CTSR*.

No.	*CTSR* Skor	STS Level
1	0–9	*Concrete*
2	10–14	*Low formal*
3	15–19	*Upper formal*
4	20–24	*Post formal*

(sumber: A.E. Lawson, *personal communication, March 19, 2014*)

2.4 *Two-tier multiple choice test of intermolecular forces*

The two-tier multiple-choice test instrument for intermolecular forces in this study is the Intermolecular Forces Diagnostic Instrument (IFDI). The development of IFDI was carried out by adopting a two-tier multiple choice instrument development procedure by Treagust (1988), whose development stages have been modified by Al-Balushi et al. (2012).

3 RESULTS AND DISCUSSION

3.1 *Student STS development level*

The level of development of STS for class XI high school students in this study is presented below.

Table 2. Development level of STS students'.

STS Level	Number of Students	%
Concrete	90	89.1
Low formal	11	10.9
Upper formal	0	0.0
Post formal	0	0.0
Total	101	100.0

Table 2 shows that many students in class XI high school experience delays in the development of their thinking skills. Several studies have shown that the existence of learning interventions in the classroom can affect the development of students' thinking skills (Adey & Shayer 1990; Effendy 1985). The influence of classroom learning on the intellectual development of students is influenced by the frequency of intellectual stimulation received by students during teaching and learning activities (Effendy 1985).

3.2 *Results of identification of students' misconceptions on intermolecular forces*

Table 3. Misconceptions of intermolecular forces identified using IFDI ($n = 69$).

	Misconceptions	Sources	% of Students Having Misconceptions
1.	Breaking of covalent bonds due to phase change	S2-C(3)	24.6
2.	Intermolecular force is stronger than intramolecular force	S5-A(2)	23.1
3.	Shape of the liquid surface	S7-B(1)	29.2
4.	Change in the volume of water in the melting and freezing processes	S8-B(1)	27.7
5.	Relationship of vapor pressure with intermolecular forces	S9-B(3)	26.2

3.3 *Ease of elimination of misconceptions for students with different STA using concept change text*

Based on the data on the distribution of individual student responses, the misconceptions per student at each STS level using IFDI is presented in Table 4.

Table 4. Number of misconceptions per student at each STS level in each IFDI item.

No.	Concept Area & Tested Concept	Item Number	Number of Misconceptions Per Student	
			C	*L*
1	The bond is broken due to a change in the phase of the substance in the molecule	2	26.8	11.1
2	Influence of intermolecular force on boiling point and melting point	5	24.1	11.1
3	Liquid surface shape	7	31.0	11.1
4	Changes in the volume of water in the melting and freezing processes	8	28.6	22.2
5	Relationship of vapor pressure with intermolecular force	9	23.2	44.4
	Rate-Rate		**26.7**	**20.0**

Description: *C* = *Concrete*; *L* = *Low Formal*

The average number of misconceptions experienced by students at each level of STS as a whole is presented in Table 5.

Table 5. Number of misconceptions that occur in students at each level of STS.

STS Level	No.	Number of Misconceptions	Rate-Rate
Concrete	60	79	1.3
Low formal	9	9	1.0

Table 5 shows that it is easy to eliminate misconceptions in students with higher level of STS. The effectiveness of learning using concept change text is also reviewed based on the percentage decrease in the number of students who experience conceptual errors after remedial learning, which is shown in Table 6.

Table 6. Percentage of reduction of student concept errors after using concept change text.

No.	Misconception	Number of Students				% Concept Error Reduction
		Misconception before Learning		Still Wrong Concept after Learning		
		Concrete	Low Formal	Concrete	Low Formal	
1	The bond is broken due to a change in the phase of the substance in the molecule	15	1	3	0	68
2	Influence of intermolecular force on boiling point and melting point	14	1	3	0	67
3	Liquid surface shape	18	1	6	0	52
4	Changes in the volume of water in the melting and freezing processes	16	2	6	0	50
5	Relationship of vapor pressure with intermolecular force	13	4	4	1	55
	Rate–rate					**58**

Based on Table 6, it is evident that the concept change text can be used as one of the methods to reduce conceptual errors in students. Pabuccu and Geban (2006) stated that concept change texts can be used to eliminate misconceptions that occur in students. Conceptual change texts make students feel dissatisfied with their conceptions and that are made specifically to make students aware that they are experiencing misconceptions, and to help students overcome misconceptions by explaining relevant examples (Hynd et al., 1994; Pabuccu & Geban, 2006).

4 CONCLUSION

Based on the results of the research, it can be concluded that: (1) 89.1% of students in class XI senior high school are at the STS concrete level and 10.9% are at the low formal level. (2) The students' misconceptions on the material of intermolecular forces are limited to: (a) there is a breaking of the covalent bond between the O and H atoms in the water molecule due to heating; (b) the intermolecular forces are stronger than the intramolecular forces; (c) the shape of a concave liquid surface is affected by the dipole–dipole intermolecular forces; (d) in the melting and freezing processes, the volumes of ice and water are same; and (e) substances having strong intermolecular forces have large vapor pressures. (3) It is easy to address misconceptions in students with higher STS levels. (4) Conceptual change text can be used to correct misconceptions in students.

REFERENCES

Adey, Philip & Shayer, Michael. 1990. Accelerating the development of formal thinking in middle and high school students. *Journal of Research in Science Teaching*, 27 (3): 267–285.

Affifanur, Yussalina. 2012. *Pengembangan dan Penggunaan Instrumen Diagnostik Two-Tier untuk Mengidentifikasi Miskonsepsi Siswa tentang Kepolaran dan Gaya Antarmolekul di SMAN 2 Jombang*. Skripsi tidak diterbitkan. Malang FMIPA UM.

Al-Balushi, Sulaiman M, Ambusaidi, Abdullan K., Al-Shuaili, Ali H., & Taylor, Neil. 2012. Omani twelfth grade students' most common misconceptions in chemistry. *Science Educational International*. 23(3): 221–24.

Barke, H-D., Al Hazari, Yitbarek, S. 2009. Misconceptions in Chemistry Addressing Perception in Chemical Education. Springer

Chandrasegaran, A. L., Treagust, D. F., & Mocerino, M. 2007. The Development of a Two-tier Multiple-choice Diagnostic Instrument for Evaluating Secondary School Students' Ability to Describe and Explain Chemical Reactions Using Multiple Levels of Representation. *Chemistry Education Research and Practice*, 8(3): 293–307.

Chittleborough G. and Treagust D. F. 2009. Why models are advantageous to learning science. *De* Anniversario. 12–17.

Coll, R. K. & Taylor, N. 2002. Mental Models in Chemistry: Senior Chemistry Students' Mental Models of Chemical Bonding. *Chemistry Education: Research and Practice in Europe*, 3: 175–184.

Effendy. 1985. *Keefektifan Pengajaran Ilmu Kimia dengan Cara Inkuiri Terbimbing dengan Cara Verifikasi Terhadap Perkembangan Intelek serta Prestasi Belajar Mahasiswa IKIP Jurusan Pendidikan Kimia Tahun Pertama.* Tesis tidak diterbitkan. Jakarta : Pasca Sarjana IKIP Jakarta

Gabel, D, L. 1999. Improving Teaching and Learning through Chemistry Education Research: A Look to the Future. *Journal of Chemical Education,* 76(4): 548–554.

Hynd, C. R., McWhorter, J. Y., Phares, V. L. & Suttles, C. W.1994. The role of instructional variables in conceptual change in high school physics topics. *Journal of Research in Science Teaching*, 31(9), 933–946.

Lawson, A. E. 1983. Predicting science achievement: The role of developmental level, disembedding ability, mental capacity, prior knowledge, and beliefs. *Journal of Research in Science Teaching*. 20(2): 117–129.

Lawson, A. E & Thompson, L. D. 1988. Formal Reasoning Ability and Misconceptions Concerning Genetic and Natural Selection. *Journal of Research in Science Teaching,* 25(9): 733–746.

Nicoll, G. 2001. A Report of Undergraduates' Bonding Misconceptions. International Journal of Science Education. 23, 707–730.

Özmen, Haluk. 2004. Some Student Misconceptions in Chemistry:A Literature Review of Chemical Bonding. Journal of Science Education and Technology, 13(2): 147–159.

Pabuccu, A. & Geban, O. 2006. Remediating misconceptions concerning chemical bonding through conceptual change text. H. U. Journal of Education, 30, 184–192.
Peterson, R. F., Treagust D. F., and Garnett, P. J. 1989. Development and application of a diagnostic instrument to evaluate grade 11 & 12 students' concepts of covalent bonding and structure after a course of instruction. Journal of Research in Science Teaching, 26, 301–314.
Piaget, J., & Inhelder, B. 1958. The Growth of Logical Thinking from Childhood to Adolescence. France: Presses Universitaires de France.
Rompayom, P., Tambunchong, C., dan Dechsri, P. 2011. Using Open-Ended Questions to Diagnose Students' Understanding of Inter- and Intramolecular Forces. US-China Education Review.
Taber, K. S., & Coll, R. 2002. Bonding. J. K. Gilbert, et al. (eds). Chemical Education: Toward Research-based Practice. 213–234.
Tan, K. D. dan Treagust, D. F. 1999. Evaluating students' understanding of chemical bonding. School Science Review, 81(294), 75–83.
Tarhan, L., Ayar-Kayali, H., Urek, R. O. dan Acar B. 2008. "Problem-based Learning in 9th Grade Chemistry Class: Intermolecular Forces". Research in Science Education. 38(3), 285–300.
Treagust, D. 1988. Development and Use of Diagnostic Tests to Evaluate Students' Misconceptions in Science. International Journal of Science Education, 10(2): 159–169.
Treagust, D. 2006. Conceptual Change as a Viable Approach to Understanding Student Learning in Science. Dalam Tobin, K. (Ed.), Teaching and Learning Science – A Handbook (hlm.25-32). Westport, Connecticut: Praeger Publications.
Tüysüz, Cengiz. 2009. Development of two-tier diagnostic instrument and assess students' understanding in chemistry. Scientific Research and Essay. Vol 4(6): 626-631
Ünal, S., Çalık, M., Ayas, A. ü., & Coll, R. K. 2006. A review of chemical bonding studies: needs, aims, methods of exploring students' conceptions, general knowledge claims and students' alternative conceptions. Research in Science & Technological Education, 24(2): 141–172.
Wang, T. & Andre, T. 1991. Conceptual change text versus traditional text and application questions versus no questions in learning about electricity. Contemporary educational psychology. 16. 103–116.
Yilmaz, D. Tekkaya, C. & Sungur, S. 2001. The comparative effect of prediction/discussion-based learning cycle, conceptual change text, and traditional instructions on student understanding of genetic. International Journal of Science education. 33(5): 607–628.

Improving Assessment and Evaluation Strategies on Online Learning – Wonorahardjo, Karmina & Habiddin (eds)

Improving the quality of online chemistry learning – A systematic literature review

I.W. Dasna* & P.A.W. Putri
Universitas Negeri Malang, Indonesia

ABSTRACT: Technological advances in the 21st century and the COVID-19 outbreak have increased the urgency of integrating digital technology in chemistry learning. This article is a literature review regarding the implementation of how online chemistry learning was accomplished during the COVID-19 pandemic. A sample of 19 articles was selected through the stages of reviewing arguments and organizing studies. Based on the results of the study, this article describes (1) the portrait of online chemistry learning during the COVID-19 pandemic, and (2) how to improve the quality of online learning. The quality of online chemistry learning enables optimal learning, an atmosphere of active learning, meaningful learning, creates good teacher and student awareness, trains the necessary skills, and maintains consistent evaluation of learning outcomes. Therefore, it needs to be continuously improved.

Keywords: Chemistry learning, organizing study material, analyzing arguments

1 INTRODUCTION

Implementation of online learning is no longer a new thing in international education. A meta-analysis regarding the implementation of online learning was conducted by the United States Department of Education, which reviewed 51 articles from 1996 to mid-2008 (Means et al. 2009); indicating that students who experienced online learning performed better than students who experienced traditional face-to-face learning. The main strength of online learning is that students can study with flexibility anywhere and anytime according to their convenience. Teaching and learning activities can be carried out by teachers and students both synchronously (same time, different places) and asynchronously (different places and times) (Brouwer & McDonnell 2009). Online learning also opens opportunities for students to get as much information as they need from the internet. In fact, the acquisition of information from various electronic devices, such as computers, smartphones, tablets, and other devices has become more popular with teachers and students than has ever been previously imagined (He et al. 2012). However, along with the popularity of online learning, there are also problems regarding how to improve and maintain the quality of learning, and carry out appropriate assessments or evaluations for these relatively new online programs (Lee-Post & Hapke 2017; Picciano et al. 2012).

Various kinds of research on online learning, especially online chemistry learning, have been conducted to determine the strengths and weaknesses experienced during learning activities. Online chemistry classes have their own strengths, challenges, and limitations. Online learning plays an important role in improving student performance, connecting junior high school students to higher education, transforming the learning environment and supporting learning cost efficiency (Picciano et al. 2012). In addition to the success of studying chemistry online, several studies have also pointed out the limitations of learning chemistry online. In particular, most traditional chemistry studies

*Corresponding Author

DOI 10.1201/9781003261346-20

spend more time doing calculations and collaborating problem-solving in the classroom, meaning students spend class time practicing higher-order thinking. Unsuccessful students spend more time on easy assignments, giving up on difficult tasks (West et al., 2006). In addition, lack of interaction with classmates, absence of teachers, and lack of feedback from classmates and teachers makes it difficult for students to stay motivated in online classes (Kress et al., 2012).

Future chemistry learning should not leave chemistry characteristics, and it is important to incorporate three important interactions in online learning, which are the interactions between teachers and students, students and students, and students with content (Marks et al. 2005). Kirtman (2009) believes that in traditional learning, students have a greater opportunity to come together and create study groups, which may be more difficult for students to do in online learning. These limitations should be pursued so that they can be minimized by quality online chemistry lessons. Online learning will continue to be developed in the 21st century, where more and more people will opt online classes (Sawang et al. 2013) increasing the urgency of doing research in the field of online learning.

The urgency as described above has been increased with the COVID-19 pandemic, which has caused new problems for education around the world. Crawford et al. conducted a study to explore the first response of various higher education institutions in 20 countries to the COVID-19 pandemic, where results show that some countries are ready for online learning and some are making preparations in a hurry due to changes that are taking place very suddenly (Crawford et al. 2020). Based on these two driving factors, online learning has become very popular in educational institutions, so it is necessary to conduct an ongoing and comprehensive examination of the readiness of students, teachers, and educational institutions (Hung et al. 2010). The novelty of this review article is to analyze how online chemistry learning was done during COVID-19 with the aim of providing insight into the implementation of online chemistry learning as well as an overview of chemistry learning in the future.

This article aims to answer the problem formulation as follows.

1) How was the implementation of online chemistry learning during the COVID-19 pandemic?
2) How to improve the quality of online chemistry learning?

2 RESEARCH METHOD

The literature review has at least three following purposes: distinguishing what has been done from what needs to be done; discovering important variables relevant to the topic; and synthesizing and gaining a new perspective (Hart 1998). According to Hart, a literature review involves three stages: (1) selecting literature sources, (2) analyzing/reviewing arguments, and (3) organizing study materials.

2.1 *Literature sources*

Journal articles as source of literature are obtained using the help of the Eric Document website (https://eric.ed.gov/) and the Publish or Perish (PoP7) application. A search on Eric's website was carried out by entering the keywords "COVID-19, online learning, chemistry", selecting articles that had been reviewed (giving a tick in the peer reviewed only column), limiting the year the article was published for the past 2 years (2020–2021), choosing the type of publication in the form of a journal article, and choosing a chemical category.

The search was deepened using the PoP7 application, searches based on Google Scholar, by entering the word "chemistry learning, COVID-19" in the title column, and "online learning, chemistry" in the keyword column, limiting the year of publication of articles from 2020 to 2021, and limiting article searches to 100 article. The display of search results is sorted by number of citations, then 10 relevant articles with the highest number of citations were selected to be part of the sample. The number of sample articles used in the review was determined after reviewing the literature argument.

2.2 *Analyzing arguments*

The three main aspects that are done at the argument review stage are describing, categorizing, and explaining (Hart 1998). The activity of describing is the first important thing that must be done, namely reading carefully the literature and then identifying the theme or variable being studied, which ones need to be used and which are not. The variables found are then grouped and compared to produce categories (categorizing activities). Both of these activities allow the researcher to explain the answers to the problem formulation properly and even broaden the horizons of the problem being studied. The 19 articles obtained from the three stages of reviewing arguments were used as samples.

2.3 *Organizing study materials*

The purpose of the study organizing activity is to carry out an analysis to make the variables in the study mapped and recorded properly. Organized study can improve understanding of the variety of variables that exist, which variables can be grouped into themes/categories and how these variables relate to one another. Analysis is by nature objective and follows defined procedures using well-known and widely validated techniques, for example by content analysis. A total of 19 sample articles were analyzed using content analysis techniques to find out the full and in-depth content of the article, which then becomes the material for discussion on how to improve online chemistry learning to have better quality.

3 RESULTS AND ANALYSIS

The main sources of literature in this review are articles published in the *Chemistry Education Research and Practice*, *Journal of Chemical Education*, *International Journal of Instruction*, *Journal of Science Education and Technology*, *Journal of Technology and Science Education*, and *Interactive Learning Environment*.

3.1 *The online chemistry learning during the COVID-19 pandemic*

The general discussion that appears in research on online chemistry learning during the COVID-19 pandemic is about how the implementation of chemistry learning must rapidly shift from traditional/offline to completely online. Researchers try to describe the difficulties faced during implementing online chemistry learning as well as some of the efforts made to overcome the difficulties or limitations faced. Students around the world generally have difficulty participating in learning chemistry online during the COVID-19 pandemic due to various factors, which are (1) difficulty accessing the internet, especially in rural areas, (2) financial difficulties in buying internet quotas, (3) students are unfamiliar with the use of online platforms, (4) too many tasks to do, (5) limitations in carrying out chemistry lessons that involve activities in the laboratory, and (6) reduced learning motivation due to boredom and mental stress (Febrianto et al. 2020; Gemmel et al. 2020; Huang 2020; Tigaa & Sonawane 2020; Villanueva et al. 2020). Not only that, the results of the assessment show that the academic achievement of students with online chemistry learning is below average, so there is need for modifications in online learning strategies (Hanson 2020).

Implementing online chemistry learning during COVID-19 pandemic requires maintaining quality of learning, one of which is by pursuing active learning by adhering to the Community of Inquiry (CoI) framework (Tan et al. 2020). CoI emphasizes that learning, both online and traditional, must have three main components: social presence, cognitive presence, and teaching presence. These three components allow online chemistry learning to enable students to interact with each other, communicate, share opinions, build knowledge, and develop their understanding through active discussion activities in online classes. Teachers must also actively provide feedback and provide service time to students who are deemed less able to take part in learning or have questions, with chat facilities outside class hours (Guo et al. 2020).

Students' perceptions in implementing online chemistry learning also varied, but more negative perceptions were found of online learning. This statement is in line with research conducted by Blizak et al., that in general students' perceptions of online learning were negative, so most opted for face-to-face traditional chemistry learning after the COVID-19 pandemic ended (Blizak et al. 2020). Likewise, Febrianto et al. examined the perceptions of students in Madura, Indonesia, which resulted in negative perception data on online chemistry learning programs (Febrianto et al., 2020). Slightly different from Huang's research, which shows that students' perceptions are quite good and satisfied with the implementation of online chemistry learning (Huang 2020). However, in all three studies, it was found that students prefer to take traditional chemistry lessons rather than online to maximize learning outcomes.

3.2 *Improving the quality of online chemistry learning*

Quality of online chemistry learning refers to important principles that must be considered during the learning design process, which are: (1) ensuring that learning is accessible, (2) showing the involvement of the thinking process, and (3) training independent learning so that students' perceptions of learning will be good (Farina & Bodzin 2018). It cannot be denied that the difficulties faced by students in implementing online chemistry learning are mostly technical, for example, difficulties in accessing the internet, the uneven distribution of high-speed internet, there are even many students who do not have smartphones and laptops (Gemmel et al. 2020; Hanson 2020; Huang 2020; Tigaa & Sonawane 2020; Villanueva et al. 2020). This situation needs to be addressed immediately by ensuring the improvement of the infrastructure needed to support the optimal implementation of online chemistry learning.

Research by Nennig shows that online chemistry learning experienced by students is more emotionally satisfying than intellectually accessible, so future online chemistry learning should pay attention to how to improve intellectual access (Nennig et al. 2020). Meng wrote in his research on how to carry out high-quality online inorganic chemistry lessons (Meng et al. 2020). Three main things that must be considered are preparing the platform to be used and adjusting the learning model so that students are comfortable learning online; then when in class, the teacher should interact both ways with students and provide feedback directly; and finally, the learning strategy is made student-centered so that even outside the classroom can encourage students to take advantage of sophisticated technology to obtain information from the literature to better understand what they have learned in class and expand their knowledge.

Online learning delivery can be varied with online flipped-classes to fully support the strengths of online and traditional learning (Ranga 2020), also by using appropriate platforms. The Edpuzzle platform is a free platform that can be used to teach chemistry online. Pulukuri & Abrams shared their experience using Edpuzzle, where teachers are supported to be able to make learning videos or edit videos obtained from YouTube with features of adding notes, adding pictures, and also inserting short questions in segments in the video (Pulukuri & Abrams 2020). The teacher can emphasize and explain by using picture features that enable the teacher to, for example, explain the mechanism of chemical reactions.

An interesting finding from the results of research that is considered capable of improving the quality of online chemistry learning is the procurement of International Massive Open Online Courses (MOOCs). The purpose of procuring MOOCs is to provide students with an overview of chemistry in everyday life, how chemistry is at the university level, as well as jobs that are relevant to chemistry to students in secondary schools (Parsons 2020).

In addition, improving the quality of online chemistry learning can also be done by increasing the integrity of the assessments used. If the assessment is based on tests, it must be ensured that the test instrument questions cannot be answered solely by relying on search engines but requires thinking and analyzing skills from students (McDowell 2020). The integrity of the assessment can also be improved by utilizing supporting platforms, for example with Respondus U and Proctor U, utilizing the Lockdown Browser feature, which requires students to turn on the camera while doing tests, as well as coding lots of questions and giving time limits (Balasubramanian et al. 2020). Bopegedera suggests using another form of assessment besides tests, by using a final project to explore more fully understanding the concepts and skills of students (Bopegedera 2020).

4 CONCLUSION

The implementation of online chemistry learning has been carried out a lot, especially in 2016–2020, not to mention due to the COVID-19 pandemic period, which suggested that all chemistry learning activities be carried out online to anticipate the spread of the virus. It is important for teachers to provide chemistry lessons online that encourage more interaction between students and teachers and students. Teachers need to be discussion facilitators, so they need to be actively involved in discussion activities, then also monitor student progress and provide consistent feedback to students (Karkar-Esperat, 2018). The process of modernizing learning should be a continuous process rather than in the form of a product of an urgent situation with hasty changes (Bernard et al. 2017) so that in the future, students' success with blended chemistry learning can be ascertained.

REFERENCES

Balasubramanian, B., Desantis, C., & Gulotta, M. (2020). Assessment à la Mode: Implementing an Adaptable Large-Scale Multivariant Online Deferred-Grade Exam for Virtual Learning. *Journal of Chemical Education*. https://doi.org/10.1021/acs.jchemed.0c00767

Blizak, D., Blizak, S., Bouchenak, O., & Yahiaoui, K. (2020). Students' Perceptions Regarding the Abrupt Transition to Online Learning During the Covid-19 Pandemic: Case of Faculty of Chemistry And Hydrocarbons at the University of Boumerdes-Algeria. *Journal of Chemical Education*, *97*(9), 2466–2471. https://doi.org/10.1021/acs.jchemed.0c00668

Bokosmaty, R., Bridgeman, A., & Muir, M. (2019). Using a Partially Flipped Learning Model to Teach First Year Undergraduate Chemistry. *Journal of Chemical Education*, *96*(4), 629–639. https://doi.org/10.1021/acs.jchemed.8b00414

Bopegedera, A. M. R. P. (2020). Using familiar and new assessment tools in physical chemistry courses during covid-19. *Journal of Chemical Education*, *97*(9), 3260–3264.

Brouwer, N., & McDonnell, C. (2009). Online support and online assessment for teaching and learning chemistry. *Innovative Methods of Teaching and Learning Chemistry in Higher Education*, 123–152. http://books.google.com/books?hl=en&lr=&id=yYLrYT26YVkC&oi=fnd&pg=PA123&dq=Online+Support+and+Online+Assessment+for+Teaching+and+Learning+Chemistry&ots=MgVX1AutCC&sig=TcvOf7y5GOpCMBgA7p-lfYtLYA4

Crawford, J., Henderson, K. B., Rudolph, J., Malkawi, B., Glowatz, M., Burton, R., Magni, P. A., & Lam, S. (2020). Journal of Applied Learning & Teaching COVID-19?: 20 countries' higher education intra-period digital pedagogy responses. *Journal of Applied Learning & Teaching*, *3*(1), 1–20.

Farina, W. J., & Bodzin, A. M. (2018). Effectiveness of an Asynchronous Online Module on University Students' Understanding of the Bohr Model of the Hydrogen Atom. *Journal of Science Education and Technology*, *27*(3), 256–269. https://doi.org/10.1007/s10956-017-9722-0

Febrianto, P. T., Mas'udah, S., & Megasari, L. A. (2020). Implementation of online learning during the covid-19 pandemic on Madura Island, Indonesia. *International Journal of Learning, Teaching and Educational Research*, *19*(8), 233–254. https://doi.org/10.26803/ijlter.19.8.13

Gemmel, P. M., Goetz, M. K., James, N. M., Jesse, K. A., & Ratliff, B. J. (2020). Collaborative Learning in Chemistry: Impact of COVID-19. *Journal of Chemical Education*, *97*(9), 2899–2904.

Guo, J., Zhu, R., Zhao, Q., Li, M., & Zhang, S. (2020). Adoption of the online platforms rain classroom and wechat for teaching organic chemistry during covid-19. *Journal of Chemical Education*, *97*(9), 3246–3250. https://doi.org/10.1021/acs.jchemed.0c00822

Hanson, R. (2020). Designing E-content for teaching basic analytical chemistry in higher education: A baseline study. *Science Education International*, *31*(1), 22–28. https://doi.org/10.33828/sei.v31.i1.3

Hart, C. (1998). The literature review in research: Releasing the social science imagination. In *Doing a Literature Review* (pp. 1–25). http://www.sjsu.edu/people/marco.meniketti/courses/ARM/s0/Literature-review-Hart.pdf
He, Y., Swenson, S., & Lents, N. (2012). Online video tutorials increase learning of difficult concepts in an undergraduate analytical chemistry course. *Journal of Chemical Education*, *89*(9), 1128–1132.
Huang, J. (2020). Successes and Challenges: Online Teaching and Learning of Chemistry in Higher Education in China in the Time of COVID-19. *Journal of Chemical Education*, *97*(9), 2810–2814.
Hung, M. L., Chou, C., Chen, C. H., & Own, Z. Y. (2010). Learner readiness for online learning: Scale development and student perceptions. *Computers and Education*, *55*(3), 1080–1090. https://doi.org/10.1016/j.compedu.2010.05.004
Karkar-Esperat, T. M. (2018). International graduate students' challenges and learning experiences in online classes. *Journal of International Students*, *8*(4), 1722–1735.
Kirtman, L. (2009). An Examination of Differences in Learning Outcomes. *Issue in Teaching Education*, *18*(2), 103–116.
Kress, H., Thering, A., Lalonde, C., Kim, S., & Cleeton, L. (2012). Faculty reflections on online course development and implementation for teacher education. *International Journal of Technology, Knowledge and Society*, *8*(1), 73–83. https://doi.org/10.18848/1832-3669/cgp/v08i01/56263
Lee-Post, A., & Hapke, H. (2017). Online learning integrity approaches: Current practices and future solutions. *Online Learning Journal*, *21*(1), 135–145. https://doi.org/10.24059/olj.v21i1.843
Marks, R. B., Sibley, S. D., & Arbaugh, J. B. (2005). A structural equation model of predictors for effective online learning. In *Journal of Management Education* (Vol. 29, Issue 4).
McDowell, S. A. C. (2020). Asynchronous Online Assessment of Physical Chemistry Concepts in the Time of COVID-19. *Journal of Chemical Education*, *97*(9), 3256–3259.
Means, B., Toyama, Y., Murphy, R., Bakia, M., & Jones, K. (2009). Evaluation of Evidence-Based Practices in Online Learning. *Structure*, 66. www.ed.gov/about/offices/list/opepd/ppss/reports.html
Meng, Y. L., Song, X. Z., Li, Y., Tan, Z., Yan, Y., & Zhang, X. (2020). High-Quality Inorganic Chemistry Teaching during COVID-19. *Journal of Chemical Education*, *97*(9), 2945–2949.
Nennig, H. T., Idárraga, K. L., Salzer, L. D., Bleske-Rechek, A., & Theisen, R. M. (2020). Comparison of student attitudes and performance in an online and a face-to-face inorganic chemistry course. *Chemistry Education Research and Practice*, *21*(1), 168–177. https://doi.org/10.1039/c9rp00112c
Parsons, A. F. (2020). Exploring Everyday Chemistry: The Effectiveness of an Organic Chemistry Massive Open Online Course as an Education and Outreach Tool. *Journal of Chemical Education*, *97*(5), 1266–1271. https://doi.org/10.1021/acs.jchemed.9b01151
Picciano, A. G., Seaman, J., Shea, P., & Swan, K. (2012). Examining the extent and nature of online learning in American K-12 Education: The research initiatives of the Alfred P. Sloan foundation. *Internet and Higher Education*, *15*(2), 127–135. https://doi.org/10.1016/j.iheduc.2011.07.004
Pulukuri, S., & Abrams, B. (2020). Incorporating an online interactive video platform to optimize active learning and improve student accountability through educational videos. *Journal of Chemical Education*. https://doi.org/10.1021/acs.jchemed.0c00855
Ranga, J. S. (2020). Online Engagement of Commuter Students in a General Chemistry Course during COVID-19. *Journal of Chemical Education*, *97*(9), 2866–2870.
Sawang, S., Newton, C., & Jamieson, K. (2013). Increasing learners' satisfaction/intention to adopt more e-learning. *Education and Training*, *55*(1), 83–105. https://doi.org/10.1108/00400911311295031
Tan, H. R., Chng, W. H., Chonardo, C., Ng, M. T. T., & Fung, F. M. (2020). How Chemists Achieve Active Learning Online during the COVID-19 Pandemic: Using the Community of Inquiry (CoI) Framework to Support Remote Teaching. *Journal of Chemical Education*, *97*(9), 2512–2518.
Tigaa, R. A., & Sonawane, S. L. (2020). An International Perspective: Teaching Chemistry and Engaging Students during the COVID-19 Pandemic. *Journal of Chemical Education*, *97*(9), 3318–3321.
Villanueva, M. E., Camilli, E., Chirillano, A. C., Cufré, J. A., De Landeta, M. C., Rigacci, L. N., Velazco, V. M., & Pighin, A. F. (2020). Teaching instrumental analytical chemistry during covid-19 times in a developing country: Asynchronous versus synchronous communication. *Journal of Chemical Education*, *97*(9), 2719–2722. https://doi.org/10.1021/acs.jchemed.0c00664
West, W., Rosser, B. R. S., Monani, S., & Gurak, L. (2006). How Learning Styles Impact E-Learning: A Case Comparative Study of Undergraduate Students Who Excelled, Passed, or Failed an Online Course in Scientific/Technical Writing. *E-Learning and Digital Media*, *3*(4), 534–543.

Improving Assessment and Evaluation Strategies on Online Learning – Wonorahardjo, Karmina & Habiddin (eds)

STEM-PBL and its effect on improving students' concept understanding in high school chemistry learning

P.A. Wulandari*
Universitas Riau, Riau, Indonesia

I.W. Dasna & Nazriati
Universitas Negeri Malang, Indonesia

ABSTRACT: The application of STEM-PBL in the world of education is inseparable from the times. This study aims to determine the effect of using STEM-PBL in chemistry subjects, especially chemical equilibrium materials. The samples were students of SMAN class XI, which consisted of one experimental class and one control class. This study uses a moderator variable, namely the initial ability of students. The study used a covariate test (two-way ANOVA). The results of the hypothesis test state that the experimental class has a significance of 0.000, which means that there are differences in students' conceptual understanding in the experimental class and the control class.

Keywords: STEM-PBL, initial ability, concept understanding

1 INTRODUCTION

The increasing focus on developing science, technology, engineering, and math (STEM) skills in students reflects the evolution of society. Along with the times, in addition to having the skills to complete work, there is also a demand to be technically competent and able to adapt to changes in work processes and the work environment. With these changes, workers need the skills to perform more cognitively demanding tasks, apply skills and knowledge to solve complex problems, and succeed in interdisciplinary and interdisciplinary teams (Heerwagen 2016).

Several countries have implemented policy reforms related to the use of STEM in learning, namely the US, Australia, Turkey, and Indonesia (Bennett et al. 2020; DENİZ & KURT 2022; LaForce et al. 2016; Sagala et al. 2019). This approach is a learning approach that can be done in various disciplines of STEM. The integration of STEM in learning is a learning program that combines two or more areas of knowledge (science, technology, engineering, mathematics) contained in it. Learning with the STEM approach is a new approach in the development of a multidisciplinary world of education (Laboy Rush 2010). An effective STEM education approach involves integrating the STEM discipline to support concept understanding and skill development in the 21st century (Carla et al. 2020).

STEM literacy, i.e., awareness of multiple disciplines, such as science, technology, engineering, and mathematics, and familiarity with key concepts in each field, should be an educational priority for all students (Bybee 2010; National Academy of Engineering dan Dewan Penelitian Nasional 2014). The STEM approach aims to develop STEM-aware students (Bybee 2010): (1) knowledge, attitudes, and skills to identify problems in life situations, to describe and design natural phenomena, and to draw evidence-based conclusions on STEM-related problems; (2) understand the characteristics of the STEM field as a form of human-initiated knowledge, research, and design; (3) awareness of how the STEM field shapes the material, intellectual, and cultural environment; (4) willingness to engage in research on STEM-related issues (e.g., energy efficiency,

*Corresponding Author

DOI 10.1201/9781003261346-21

environmental quality, shortage of natural resources) as a constructive citizen; use, nurture, and think about ideas from technology, engineering, and mathematics.

The STEM approach emerges in certain situations where learning science or mathematics involves full-scale problem-solving activities in social, cultural, and functional contexts. The application of STEM can be supported by various learning methods. STEM is comprehensive and can support its implementation by using various learning models. STEM Education Policy Documents and literature review recommend a research-based approach (Bybee & Fuchs 2006; National Research Council 2000; National Research Council 2012; National Science Teachers Association 2011). One of the learning models that can be used is problem-based learning (PBL). It presents real and meaningful problem situations that can stimulate students' curiosity to explore contextual problems related to everyday life and solve them through the stages of the scientific method (Arends 2012). This is in accordance with the statement of Matthews (2007) which explains that PBL is learning that results from the learning process so that it is able to understand and solve existing problems.

Conceptual understanding, which allows children to understand ideas in a transferable way, helps students take what they have learned in class and apply it to an entire topic. Government accountability and test-measurement instruction are common, but these methods do not necessarily require students to acquire the skills to complete assignments outside of class. In a study conducted by Brown and Kane (1989), preschoolers tended to transfer skills between situations when asked to use previously taught and proven solutions. They learn best when they see a model solution, rather than being given explicit rules. In this sense, children in scientific disciplines are based on new understanding gained by observing examples of solutions, rather than clear rules that represent problems or possibilities for solving problems. You must be able to make decisions.

STEM-PBL has been widely used in school subjects, including chemistry. The many benefits that can be obtained from learning using STEM-PBL, researchers feel it is important to conduct research related to STEM-PBL in chemistry subjects. In this study, the effect of using STEM-PBL to improve student's concept on chemical equilibrium materials will be seen.

2 METHODS

Based on the objectives to be achieved, the research design used in this study was experimental, a pure post-test design using a 2 x 2 factor experiment. The 2 x 2 factor experiment is shown in Table 1. The sampling method uses a targeted random sample (Creswell 2012). The samples used were two classes, one as the experimental class and the other as the control class. The sample used is class XI SMAN students in Malang Regency.

Table 1. Research design quasi experimental factorial 2 x 2.

	Learning model	
Initial ability	PBL + STEM (X_1)	PBL (X_2)
High (Y_1)	X_1Y_1	X_2Y_1
Low (Y_2)	X_1Y_2	X_2Y_2

Adapted from (Fraenkel, J.R, Norman F Wallen, 2012)
Description:
X_1Y_1 = understanding concepts with high initial abilities taught by PBL-STEM
X_1Y_2 = understanding concepts with high initial abilities taught by PBL
X_2Y_1 = understanding of concepts with low initial abilities taught by PBL-STEM
X_2Y_1 = understanding of concepts with low initial abilities taught by PBL

This study used two types, namely treatment instruments and measurement instruments. The treatment instrument consisted of (1) lesson plans (2) worksheets while the measuring instrument was a concept-understanding test. The results of the validation of the lesson plan and student worksheet treatment instruments were 0.89 and included in the high category. Initial ability is used as a moderating variable, taking into account the potential impact on students' conceptual understanding and reasoning abilities. As a preparatory step, a prerequisite test and an average similarity test (t-test) were carried out on the initial ability data of students in the two classes. Based on the initial ability data obtained, students are further divided into students with higher initial abilities and students with lower initial abilities. Analysis of learning outcomes data in the form of concept understanding scores and reasoning abilities was carried out in two stages, namely prerequisite tests (consisting of normality test and homogeneity test) and hypothesis testing.

3 FINDINGS

3.1 *Students' initial ability*

To prove that the average initial ability level of students in the two classes is the same, proceed with a preliminary test (homogeneity test, normality test) followed by an average similarity test. The Kolmogorov-Smirnov test used for normality analysis obtained a significant value for the experimental class of 0.2 and the control class of 0.148, which indicates that the initial ability in both classes is normally distributed. In addition, a prerequisite test in the form of Leven Test was used to analyze the homogeneity of the data of students' initial abilities, which obtained a significance value of 0.561 so that it can be concluded that the data of students' initial abilities were in the control and experimental classes are homogeneous. Furthermore, the t-test (Independent Sample T-Test), which was conducted to test the average similarity in the two classes concluded that there was no difference in the initial abilities of the control class students and the experimental class students' initial abilities.

3.2 *Student concept understanding*

Students' conceptual understanding was measured using a two-tier objective test with five alternative answer choices. The data on the acquisition of students' conceptual understanding in the experimental class and control class can be seen in Table 2. Based on the results obtained, it can be seen that the average value of students' conceptual understanding in the experimental class is 80.39 and in the control class is 69.69.

Table 2. Distribution of student concept understanding.

Class	Amount of student	Average	Max	Min
Experiment	28	80.39	98	65
Control	29	69.69	90	52

Students' understanding of concepts is categorized based on students' initial abilities. Each class, both experimental class and control class, are grouped into students' conceptual understanding with high initial ability and concept understanding students with low initial ability. The grouping of concept understanding based on this initial ability can be seen in Table 3.

Table 3. Understanding student concepts based on students' initial ability.

Class	Group of Concept Understanding	Amount	Average	Max	Min
Experiment	High	14	81.07	98	68
	Low	14	70.71	95	65
Control	High	15	73.53	90	52
	Low	14	64	80	53

Based on the prerequisite tests that have been carried out on both classes, namely the experimental class and the control class, the results show that the two classes are normally distributed and homogeneous. Hypothesis testing was carried out using the Two Way ANOVA method of analysis of variance assisted by the SPSS 22 for Windows application with a significance level of 0.05. The results of hypothesis testing students' understanding of concepts can be seen in Table 4.

Table 4. Hypothesis testing analysis of variance on concept understanding.

Dependent Variable: Concept Understanding					
Source	Type III Sum of Squares	Df	Mean Square	F	Sig.
Corrected Model	2508.469[a]	3	836.156	9.555	.000
Intercept	318835.666	1	318835.666	3643.288	.000
Class	1578.929	1	1578.929	18.042	.000
Concept_Understanding	589.250	1	589.250	6.733	.012
Class * Concept_Understanding	382.512	1	382.512	4.371	.041
Error	4638.198	53	87.513		
Total	326348.250	57			
Corrected Total	7146.667	56			

[a]R Squared = .351 (Adjusted R Squared = .314)

4 DISCUSSIONS

The results of the analysis of the average values in Table 2 prove that there are differences in students' understanding of concepts taught using the PBL-STEM model and the PBL model. The average value of students' understanding of concepts taught using the PBL-STEM model is 80.39. This value is higher when compared to the average value of understanding the concept of students who are taught using the PBL model is 69.69. The results of the two-way ANOVA hypothesis test in Table 3 also show that there are differences in students' understanding of concepts taught using the PBL-STEM model and the PBL model. This shows that students who are taught using the PBL-STEM model have a better understanding of concepts when compared to students who are taught using the PBL model. The results of the study showed that the combination of the PBL learning model with the STEM approach was able to improve a better understanding of concepts when compared to using only the PBL learning model.

Learning activities between classes taught with the PBL-STEM model and PBL in general are not much different. Both classes were taught according to the learning syntax using the PBL model. The difference between these two classes lies in the given STEM approach. In the experimental class, students are given more complex problems, namely the provision of problems related to industrial technology that exists in real life. When faced with complex problems, students will be required to have a high conceptual understanding because it requires the involvement of many concepts in solving them. In one of the learning activities in the experimental class, a case study was given about the use of N2O4 as the main oxidizer of a rocket, students were asked to solve a problem related to how to keep the balance of the N2O4 contained in the rocket not turning into NO2. In the control class, a problem is given about how the equilibrium occurs in the decomposition reaction of N2O4 into NO2.

The results of the study showed that the combination of the PBL learning model with the STEM approach was able to improve a better understanding of concepts when compared to using only the PBL learning model. This is because STEM plays a role in providing more complex problems that demand a better understanding of student concepts.

The initial abilities possessed by students have an influence on students' understanding of concepts. Students with high initial abilities tend to have a better understanding of concepts than students with low initial abilities (Danial et al. 2017; Hasrida et al. 2018). Concept understanding of students with different initial abilities can be seen in Table 2. In the PBL-STEM class, students who have high initial ability have a higher average concept understanding compared to students who have low initial ability (81.07 > 73.53). This also applies to the PBL class where students with high abilities have a higher average understanding of concepts than students who have low initial abilities (70.71 > 64.00).

Based on the results of understanding students' concepts in both the experimental class and control class, it shows that the initial ability of students is one of the factors that play a role so that students are able to understand the concept of chemical equilibrium. Students' initial ability is measured through a test before the equilibrium material is given, namely the reaction rate material, which is a prerequisite for studying chemical equilibrium material. The results of the initial ability test show that students who have high initial abilities in the reaction rate material are also able to understand concepts well in the chemical equilibrium material. Students tend to be more prepared and superior in accepting concepts in chemical equilibrium material when they have high initial abilities than students who have low initial abilities.

5 CONCLUSIONS

Based on the results of research that has been carried out, namely by looking at the effect of STEM-PBL learning on chemical balance subjects, it can be seen that there is a significant effect on increasing student understanding when taught using STEM-PBL. In addition, students' initial abilities also have an influence on students' understanding of concepts. Other subjects in school can be taught to students using STEM-PBL learning.

REFERENCES

Arends, Richard. (2012). Learning to Teach 9th Edition. New York: McGraw-Hill.

Bennett, D., Knight, E., Dockery, A. M., & Bawa, S. (2020). Pedagogies for employability: understanding the needs of STEM students through a new approach to employability development. Higher Education Pedagogies, 5(1), 340–359. https://doi.org/10.1080/23752696.2020.1847162

Becker, K. and Park, K. (2011). Effects of integrative approaches among science, technology, engineering, and mathematics (STEM) subjects on students' learning: A preliminary meta-analysis. Journal of STEM Education: Innovations and Research, 12(5/6), 23–37.

Bybee, R. W. (2010). Advancing STEM education: A 2020 vision. Technology and Engineering Teacher, 70(1), 30–35.

Carla C. Johnson, et al. (2020). Handbook of Research on STEM Education. New York. Rotlege. https://doi.org/10.4324/9780429021381

Creswell, J. W. (2012). Educational Research?; Planning, Conducting and Evaluating Quantitative and Qualitative Research (Fourth Edi). United States of Amerika: Pearson Education.

Cross, D. (2007). Creating Optimal Mathematics Learning Environments: Combining Argumentation and Writing to Enhance Achivement. Disertasi. University of Georgia: Tidak diterbitkanCreswell, J. W. (2012). Educational Research?; Planning, Conducting and Evaluating Quantitative and Qualitative Research (Fourth Edi). United States of Amerika: Pearson Education.

Danial, M., Gani, T., & Husnaeni, H. (2017). Pengaruh model pembelajaran dan kemampuan awal terhadap kemampuan berpikir kritis dan pemahaman konsep peserta didik. Journal of Educational Science and Technology (EST), 3(1), 18. https://doi.org/10.26858/est.v3i1.3509

DENİZ, Ş., & KURT, G. (2022). Investigation of Mathematical Modeling Processes of Middle School Students in Model-Eliciting Activities (MEAs): A STEM Approach. Participatory Educational Research, 9(2), 150–177. https://doi.org/10.17275/per.22.34.9.2

Fraenkel, J.R, Norman F Wallen, H. H. H. (2012). How to Design and Evaluate Research in Education (8th ed.). New York: McGraw-Hill.

Hasrida, Muhammad Danial, P. S. (2018). Pengaruh Model Pembelajaran dan Kemampuan Awal Terhadap Pemahaman Konsep dan Motivasi Belajar. 2(2), 44–57.

Laboy Rush, D. (2010). Integrated STEM Education through Project-Based Learning. New York: Learning.com.

LaForce, M., Noble, E., King, H., Century, J., Blackwell, C., Holt, S., …Loo, S. (2016). The eight essential elements of inclusive STEM high schools. International Journal of STEM Education, 3(1). https://doi.org/10.1186/s40594-016-0054-z

Mattews-Aydinli, J. (2007). Problem based learning and adult English language learners, Center for Adult English language acquisition, Center for Applied Linguistics. (http://www.cal.org/caela/esl_resources/briefs/Problem-based.pdf)

National Academy of Engineering and National Research Council. (2014). STEM Integration in K-12 Education: Status, Prospects, and an Agenda for Research. Washington, DC: The National Academies Press.

Sagala, R., Umam, R., Thahir, A., Saregar, A., & Wardani, I. (2019). The effectiveness of stem-based on gender differences: The impact of physics concept understanding. European Journal of Educational Research, 8(3), 753–761. https://doi.org/10.12973/eu-jer.8.3.753

Improving Assessment and Evaluation Strategies on Online Learning – Wonorahardjo, Karmina & Habiddin (eds)
© 2022 copyright the Editor(s), ISBN: 9781032198903
Open Access: www.taylorfrancis.com, CC BY-NC-ND 4.0 license

Author index

Affriyenni, Y. 23, 89
Afifah, N.N. 49
Amin, M. 75
Andayani, S. 63
Ardyansyah, A. 37
Ariani, N. 56

Dasna, I.W. 129, 135

Ediyanto, E. 1, 8
Efendi, M. 1, 8
Effendy 123
Eva, N. 63

Fadilah, N. 115
Fauzan, L. 75
Febriana, A. 108
Febrianti, Y. 56

Habiddin 37
Hakiki, R. 108
Hanafi, H. 75
Hartanto, D. 8
Hayashi, Y. 101
Hidayah, N. 75
Hirashima, T. 101
Hitipeuw, I. 1

Joharmawan, R. 108
Junaidi, A.R. 8

Karmina, S. 82
Kenila, E. 1
Kharismaliyansari, K. 23

Lestari, B.I. 82

Mawangi, P.A.N. 96
Ma'rufah, D.M. 123
Muchson, M. 108
Mukhlis, I. 44
Munzil 23

Nazriati 135

Parlan 37
Pinandito, A. 101
Pramudhita, A.N. 96
Prasetya, D.D. 101
Prihandini, T.F. 16
Putra, A.B.N.R. 115
Putri, P.A.W. 129

Rahayu, D.E. 115
Rahmawati, F. 44
Ramli, M. 75
Rini, A.M. 115
Rini, D.R. 16
Rokhim, D.A. 37
Rosli, M.S. 68

Setiawan, N.C.E. 37, 68
Setyowati, L. 82
Sujiatmoko, A.H. 82
Sukmasari, T. 115
Sunandar, A. 8
Sutadji, E. 115
Sutopo, S. 89
Syafruddin, A.B. 37

Widarti, H.R. 37
Wijaya, H.E. 49
Wisesa, A.M. 16
Wonorahardjo, S. 123
Wulandari, P.A. 135
Wulandari, R.T. 16

Xiong, W. 82

Yahaya, M.F. 29
Yogihati, C.I. 89
Yoto 96
Yuniarti, N. 8
Yusida, E. 44